# THE
# EIGHT
# MASTER
# LESSONS
# OF
# NATURE

ALSO BY GARY FERGUSON

*Land on Fire*
*Hawk's Rest*
*Opening Doors*
*The Carry Home*
*Through the Woods*
*Decade of the Wolf*
*Nature's Keeper*
*The Great Divide*
*Shouting at the Sky*
*Spirits of the Wild*
*Yellowstone Wolves*
*Walking Down the Wild*

# THE
# EIGHT
# MASTER
# LESSONS
# OF
# NATURE

*What Nature Teaches Us*
*About Living Well in the World*

# GARY FERGUSON

**DUTTON**

**DUTTON**

An imprint of Penguin Random House LLC
penguinrandomhouse.com

LIBRARY OF CONGRESS CATALOGING-IN-PUBLICATION DATA
Names: Ferguson, Gary, 1956– author.
Title: The eight master lessons of nature : what nature teaches us about
living well in the world / Gary Ferguson.
Description: New York, New York : Dutton, [2019]
Identifiers: LCCN 2019001906 | ISBN 9781524743383 (hardcover) |
ISBN 9781524743406 (ebook)
Subjects: LCSH: Nature.
Classification: LCC QH81 .F377 2019 | DDC 508—dc23
LC record available at https://lccn.loc.gov/2019001906

Printed in Canada
1   3   5   7   9   10   8   6   4   2

Book design by Nancy Resnick

*For my dear Mary,*
*who shows me every day what it looks like to seek—and to find—*
*the bright and reliable goodness in us all*

# Contents

# CONTENTS

# Coming Home

There was a time long ago when the supple reach of trees and the whisper of water bore deep comfort. When the ringing of birds and the mutter of frogs reminded us to take heart in the fact that whatever else was happening, we were all still here, rising strong in the morning, faces to the sun. We knew the juice of spring then, surging in the wild rose and the fig, and in us too—we could feel it, as if we were somehow wed to the tender shoots, as if our own breathing was the great green breath of Earth. A time when a person barely knew where she ended and the world began. The shouts, the songs, the purring and the whirring—all of it an astonishing conversation.

From where we stand now, it can seem like a dream. But every so often, out of nowhere, the thrill of that old dance judders in our bones. It reawakens sensations we had within easy reach when we were young, still wide-eyed for the world.

My own inklings of such things came during a boyhood spent in the Midwest, in a small city in northern Indiana nursed on corn and casseroles and steel. A place where nature got served in teaspoons: A garden behind our house about the size of two couches laid end to end, where I first stood barefoot and watched the comings and goings of butterflies and bees. A slim ribbon of grass where in the twilight of summer I scurried about with a jam jar, scooping up lightning bugs. And the sidewalks of Twenty-Seventh Street—walking home from school on some gloomy April afternoon with the wind in the maples, thrilling to big booms of thunder thunking inside my chest.

Around age eleven I started riding my purple Stingray ten blocks west to Potawatomi Park, a place laced with lines of oaks and maples with trunks so big I couldn't reach even halfway around them. And nearby, a greenhouse with a banana tree and orchids and an avocado tree. And next to that, a pocket zoo with a geriatric lion, six crowing peacocks, a display of roaches (really), and a snorting donkey.

As I have in adulthood spent some forty years waltzing through some of the world's wildest places, drifting along more than thirty thousand miles of trail through the outback, you might imagine I'd think of these modest encounters with nature when I was young as, well, quaint. But no. They were big: in their own way as vital as the yawning savannahs of East Africa or the high, lonesome roll of the

Yellowstone backcountry or the cold, quiet reaches of the Arctic tundra.

The lightning bugs and the bees and the giant oaks at Potawatomi Park are what cracked me open, what introduced me to the ebb and flow of the world, to the daily playlist of chirps and buzzes and snorts and whooshes. Not just piquing my curiosity, but also leaving me with the extraordinary sensation of being a part of it all. The chrysanthemums in the flower box under my bedroom window were somehow set to my own bright sense of summer. The cardinal in our maple tree wore a dazzling feathered suit that was my color of red. The mourning dove in the apple tree beside Carl and Yvonne Wilson's driveway cooed a song that even on first hearing was somehow familiar, reassuring.

No matter who you are, or where as a child you happened to live, more than likely you, too, took some of your first steps toward growing up and growing out because nature charmed you into the precious work of unfolding. Whether it comes in big helpings of the wild, or as a spider web in the corner of your garage, or even as dandelions growing out of the cracks in the sidewalk, nature calls to each of us.

And here's the thing: Despite what you might have been led to think, such magic doesn't just disappear, quietly drifting forever out of reach. After all, as we go through the various stages of our lives, we keep *adding to* what's come before, rather than replacing it. Wherever you are now, however urban or interior your life, nature is still there for you—anchoring, inspiring, helping you become more of what it is you set out to be. At the same time, the natural world remains a ready source of essential lessons, each one helping us better understand what life really needs in order to thrive.

Much of what follows is about catching up with what science is uncovering daily about how nature really works—and how you work too—using these discoveries to help us more fully comprehend what it means to be alive in this world. But these pages are also about befriending the powerful emotions that nature often ignites in us—trusting them to guide us into the deepest, most satisfying parts of our being.

In the late 1990s I had the great fortune of meeting a quiet, graceful, intensely curious sixty-five-year-old man from Utah named Lavoy Tolbert. Lavoy was formerly a science teacher—awesomely popular with his students, many of whom are still in touch with him today. He spent his whole life roaming the wilds of the American Southwest. Even now, at eighty-five, he remains totally wed to the ground underfoot, spending nearly as many nights sleeping on a pad under the stars as on the mattress in his bedroom. During our time together I've heard him talk often about how for thousands of years people have gone to nature to figure things out, to make adjustments to how they understand the world, and from that, to refine how they live. So, too, for Lavoy. He says it's a case of credentials.

"Think about it. Out in nature you've got 4.6 billion years of success—the absolute best of everything. The finest the world has come up with, all around you, night and day. Go out for a stroll in the woods and you walk among champions!"

It's kind of like this farmer, he likes to tell me. Every week, month after month, year after year, the farmer takes his old plow horse to the racetrack to run against the thoroughbreds. One day a friend stops him, asks why he keeps paying entrance fees to run races he knows he'll never win.

"You're right about that," says the farmer, rubbing his chin. "That old horse ain't got a chance. But then he sure does like the company."

In the pages that follow, you too will be in some extremely good company: pieces of nature big and small—in the yard, in the park, in wild places, but also now and then hanging out with that vast web of nature inside of *you,* a being who in truth is another magnificent example of the power and grace of Earth's creative forces.

As much as anything, this is a book about coming home. Not just back to where you grew up—not only to the suburban split-level with the clipped lawns and the cement birdbath; not just back to the brick apartment building or the small white Craftsman house on a tree-lined street in the heart of the city. Not only back to the old farm with the barn and the green tractor and the moon-eyed cows. No, this bigger, deeper home has to do with the profound belonging that comes from recovering your natural camaraderie with the wonder of the planet. And then learning to use that camaraderie as a guide for your life.

On those days when we're overly busy—and right now it seems there are plenty of them—we can be hard-pressed to find time even to stop and turn our faces to the sky. Or notice the smell of rain. Or savor for a few seconds the sound of geese overhead, urging one another toward winter grounds. But here's something to ponder: The extent to which we've actually lost touch with the wisdom of nature may not be just about our being distracted. It's also about *how we've been taught to think.* We are like people throughout history, in that the lion's share of our beliefs about how the world works has been cobbled together from cultural views and social mores, including scientific ones, passed down across the generations. When it comes to the nature of nature, and our relationship with the world around us, we've accepted an awful lot as being "just the way things are." And in fact they're nothing of the sort.

For example, one of the main currents—a kind of riptide, really—clutching at us for some two thousand years, is the idea that as humans we stand outside of nature. Above it. From that locus it's all too easy to end up passive toward, disconnected from, even bored by, the Earth.

At the same time maybe we've joined that group of resolute folks who long ago set off to master a fixed set of immutable laws of the physical universe, the goal being to gain control over what can be a maddeningly unpredictable life on a maddeningly unpredictable planet. And of course to no small extent we *have* gained measures of control. Our cleverness has given us mountains of useful things—from lasers to medicine, fast cars to cell phones, movies and blue jeans and rockets to Mars. Yet along the way we've partitioned and quarantined life—including ourselves—breaking it into smaller and smaller pieces because we think in the end this will yield certainty.

These worldviews aren't ultimate realities, but rather single strands of perception. Which means we can add to them. Growing our connections to nature will allow us to push past mere intellect to embrace sensory experiences, emotions, intuition. Of course we can continue to parse, to disassemble, to make predictions. But all that becomes more rightsize when we celebrate how much of life is mysterious, beyond anyone's capacity to understand. Curiously, while we often think of science as all about answers, the science of today is very much about living with questions. For the first time, a large body of research is illuminating the fact that our world arises from a sprawling, highly dynamic set of rhythms and relationships. It's becoming ever more clear that in the truest sense there's really no such thing as a tree, a dog, a sunflower, a human. At least not in the way we've long thought of them, as stand-alone beings. We shape—and are thoroughly shaped by—the vast array of life-forms and processes with which we share this planet.

An especially intriguing aspect of this scientific shift—of this effort to break science out of the intellectual boxes we built for it—is the effort by some to merge traditional science with indigenous wisdom. Harvard ethnobotanist

Shawn Sigstedt, for example, was, in the 1990s, one of a growing number of non-native researchers to begin working alongside native scientists, including celebrated Hopi tribal member Frank Dukepoo, a professor of genetics at Northern Arizona University. Sigstedt said the collaboration expanded his perspective in a way that helped him see the world through the lens of connection.

"Traditional culture has helped us realize our blind spots," Sigstedt noted. "The Native understanding of the world—as a place of process and relationship—is completely different than our own." As a result, Sigstedt began designing research around very different questions.

And by changing what we ask, we change the world.

As MIT systems scientist and bestselling author Peter Senge points out, a key effect of our having lost touch with the totality of nature is that we've lost the ability to perceive *interdependence*. Most of us just don't see it.

Embracing interdependence lets us move beyond the knowledge of what makes something tick into how that something is able to keep on ticking. And that, as science is now showing us, always comes down to connections: from the fungi in the forest that nutrify the soil to the trees that sprout and grow from that nitrogen to those trees then exuding the very oxygen that allows you to find your next breath.

A number of years ago, while teaching a nature writing class in Yellowstone National Park, I had the great pleasure of having as one of my students Sister Helen Prejean. By the time we met, Sister Helen had already gained acclaim for her work helping the severely troubled and down-and-out, including her correspondence with two convicted murderers, Elmo Sonnier and Robert Willie. Those conversations ended up being the genesis for her remarkable book, and later the movie, *Dead Man Walking.*

On our final day together, the group was out hiking along the Specimen Ridge Trail toward Amethyst Mountain, the land before us awash in glorious views of the Serengeti-like grasslands of Yellowstone's Lamar Valley. Clusters of bison milled about the valley floor. Small groups of pronghorn rested along the edges of the aspen woods, while overhead red-tailed hawks pirouetted on the wind. At one point, pausing to savor the view, Sister Helen brought up the religious concept of annunciation. I'd only ever heard the word used in doctrinal terms, as in the Christian celebration of the angel Gabriel telling Mary she was destined to give birth to the Son of God. But Sister Prejean had a much broader take. She said that to her, annunciation was the act of making something fresh, of bringing an ideal more fully to life by making it real in our everyday world.

Held in her take on annunciation lies a beautiful opportunity to respect the planet in the way the ancient Greeks defined respect—which was to "look again." Indeed the time is very much here to look again—and in doing so, to begin putting the world back together.

I once stumbled across a story from the early 1920s about a young anthropologist sent west to California by Harvard University, there to begin chronicling the lives of the Pit River Indians—a culture hovering on the brink of extinction. Over months he recorded their language, learned whatever stories they were willing to share, and made extensive notes about many of their social customs. At one point, sitting in the sagebrush at the edge of a village with a group of tribal elders, he asked about their word for newcomers to the land—people like his own family and colleagues, descendants of Anglos from Europe and England.

He later recalled how the tribal leaders looked at one another, shook their heads, refused to answer. Finally, after a lot of cajoling, one of the old men took a deep breath and began to speak:

"Our word for your people is *inalladui,*" he said. "*Inalladui.*"

It's easy to imagine the young anthropologist repeating the word, savoring the liquid sound of it.

"What a lovely expression," he might have offered. Which would've surely left the elders rolling their eyes.

"It means tramp," the old man continued. "Someone not at home in the world. Your people move across the land in such a hurry. You have no interest in making connections with the animals or the plants or the people who live there. This we cannot understand. We think a part of you must be dead inside."

In a sense the Pit River Indians were right: For a very long time a part of us has been dead inside. Or at least dead to many of the essential energies that sustain our mental and physical health. As Jane Goodall observed, "We seem to have lost the connection between our clever brains and our hearts." The task isn't to strive for something brand-new. After all, where would this brand-new something come from? Rather it's time to wake up the tissues of perception that have been there all along.

We *are* nature.

When we stand firm on that undeniable fact, shedding the long-standing illusion that there's nature "out there" and then there's us "in here," we'll be able to see in a new light some of our most troubling, persistent problems. And at the same time, just as important, we'll find assurance that on the very deepest level we have absolutely everything we need.

It really is possible to mend our relationship to the world around us and, through that mending, release an intelligence millions of years in the making. The journey begins with eight lessons—each one a window into seeing both outward and inward at the same time, views that afford a very different reality than the one we settled for all those years ago.

# Mystery: Wisdom Begins When We Embrace All That We Don't Know

As we acquire more knowledge, things do
not become more comprehensible, but more
mysterious.

—Albert Schweitzer

When Albert Einstein got stuck on a problem—and
truth be told, he got stuck quite a bit—he'd often go
outside. Not to some remote wild landscape but to a little
patch of forest on the Princeton campus, maintained especially for him, known as the Institute Woods. You might
assume he was merely trying to clear his head, as a lot of us
do when we step out for a quick change of scenery. But it's
a more intriguing story.

Once out in those familiar woods, Einstein was said to
stop and look around, taking in the trees and shrubs, the sky
overhead, and the grasses underfoot. At first he'd try to

imagine the workings of it all, knowing full well he couldn't do it. Consider that even today, more than sixty years after his death, we still don't fully understand everything that's happening in a square yard of dirt, let alone a patch of woods. But that was the point: He wanted to *intentionally* overwhelm himself. Get disoriented. Blow his mind. And thus, with his intellect brought to its knees, Einstein consistently found himself in a freer, more intuitive space.

Look deep into the mysteries of nature, he liked to say, and then you will understand better.

Einstein, like many other great scientists, knew that no problem was ever solved on the plane where it first revealed itself. So he used the woods to lift himself to a higher place, one less defined and more creative. Touching the considerable mystery afforded by that modest grove of trees allowed him to connect with what he considered "the source of all true art and science." He said as much to his students too, advising them that if they had a choice between gaining knowledge and maintaining a relationship with mystery, they should choose mystery.

That kind of choosing requires a very different kind of intelligence than the one with which the culture of Einstein's time was familiar, or even comfortable talking about. Yet he was absolutely certain that those who wouldn't or couldn't connect with mystery were "if not dead, then at least blind."

Albert Einstein isn't the only superstar with a penchant for dialing into mystery. Peering into the heavens at night,

Carl Sagan claimed that science wasn't only compatible with mystery but was a profound source of it. The mystery that's revealed, he said, "when we recognize our place in an immensity of light-years and in the passage of ages, when we grasp the intricacy, beauty, and subtlety of life . . . is surely spiritual."

Contemporary physicist Edward Witten, long a champion of string theory and arguably one of the smartest humans on the planet, sees mystery at the most fundamental layer of human existence. Meanwhile, Jane Goodall remains unwilling to explain life through truth and science alone. "There's so much mystery. There's so much awe."

If we're to make friends again with mystery, we'd be smart to learn something about where it likes to show up. One place it feels totally at home, of course, is in the arms of wonder. Which is lucky, because in truth we live in a time of extraordinary "scientific wow!" Who wouldn't be a little thrilled to learn that spiders can fly by employing electrical charges in the atmosphere? Standing on their hind legs, they cast silk into the air. That silk is negatively charged and repels similar negative charges in the surrounding atmosphere, sending the spiders ballooning into the heavens. Or who wouldn't feel a twitch of bewilderment to think that 99.99999 percent of our body is comprised of the empty space that exists between the electrons, neutrons, and protons—each one of those an element of the atoms that give us form. Furthermore, if you got rid of all this space, then the actual mass of your body—your "substance"—would be so small you couldn't even see it. In fact, if we took away all the space in all the bodies of every human being on the planet, the mass that remained would be about the size of a sugar cube.

Think, too, for a minute about the fact that as you walk down the street today you won't really be making contact with the ground. Rather, the magnetic force of the electrons in your shoes will be pushing away the electrons in the pavement, which means that at a supremely close-up level you really aren't walking through your life with your feet on the ground at all. You're floating. And then, of course, that old favorite: the fact that you could blast off from Earth on a journey to find the end of space, travel a hundred thousand miles an hour for the next ten thousand years, and not be one inch closer.

Or how about the extraordinarily curious fact that any of us is here at all? If the gravity in the universe had been stronger by just a tiny bit, then the stars that formed because of that gravity would have been much smaller. Our own sun might have lasted for only ten or twenty thousand years, fizzling out long before we humans or any other creature had a chance at life. Likewise, if the strength of the force that binds the nucleus of an atom were just a wee bit weaker, there wouldn't be the current range of chemicals in the universe. And without that perfect array of chemical complexity: no life.

Far from disenchanting nature, as early scientists of the seventeenth century pledged to do, contemporary biology, physics, medicine, and ecology are admitting its enchantment all over again, inviting us with every passing year to perceive a more unbridled universe. In recent decades especially, scientists have found themselves looking beyond the desire to hold reality still, drawn instead to the mysterious and fluid nature of the planet and the life it sustains. Showing us with every passing day how the physical and even psychological functions of this life are vigorous, dynamic, and effervescent.

The discoveries of modern science are more than fantastic enough to explode our intellects, which is really the first order of business when it comes to befriending mystery. But you can't really see mystery face on. It takes looking sideways, like the way you can only see certain faint stars by glancing slightly off center. The trick, in other words, is to manage the tools of perception.

Our perception expands when we realize that nature is engaged in a very big game of passing things back and forth—a mysterious ebbing and flowing, an appearing and disappearing act once described by philosopher Neil Evernden as the rhythm of exchange. One thing talks while another listens. One thing touches ground while another lifts off and flies. One part of the system waxes while another wanes. One thing dies and another is born.

When we link this rhythm-of-exchange idea to the concept of nature in ancient Greece—*phusis,* which refers to nature as life emerging from itself—what begins to shimmer on the edges of our imaginations is an unflagging set of symphonic movements all linked together to create the vast, incomprehensible weave that both holds and extends far beyond the more obvious anchor points of daily life. In truth, your very existence depends not so much on "things" as on rhythms of relationship that transcend the boundaries of your skin. It's a mind-boggling set of interactions that renders everything inside you and around you different this day than the last, all of it shifting and emerging and passing away.

So if this is how the world is—such a grand river that you can't immerse your foot in the same place twice—how come we have such a hard time seeing it that way? For starters, consider that while Albert Einstein, Jane Goodall, and Carl Sagan were building their perception and ultimately their inspiration from what can't really be known, embracing mystery at every turn, key parts of society were preoccupied with essentially organizing the closet. And our educational systems in particular were being reworked by folks with some of the tidiest closets of them all. Too often, schools clipped the wings of wonder in our children and taught them instead to regurgitate facts. In musical terms, you might say the school orchestra handed us a stick and a plastic tub to beat on, when what we were born for is a grand piano.

To be clear, education wasn't always like this. There

were times when we acted differently, when we chose to strengthen that wonder. Like the enthusiasm in the United States between roughly 1910 and 1920 for planting school gardens. While the movement was prompted in part by people leaving farms and taking up lives in the city, it had to do with a lot more than making kids aware of where their food came from. Much like the modern school garden movement of our own time, gardens back then were seen as a way to build on children's natural connections to nature, using that as a baseline for nourishing curiosity. And curiosity was considered essential for fostering critical thinking skills.

To be sure, over time a child might well come to see why a butterfly visits only certain flowers in the garden, grasping that it's the only insect with a proboscis long enough to reach the nectar. But long before that there would be the butterfly itself, dancing on velvety wings the color of twilight and autumn leaves, igniting an eagerness to look deeper—and by looking deeper, to begin to learn. As biologist Rachel Carson said, when it comes to guiding children, it would help to remember that it isn't half so important for them to know as to *feel*.

"If facts are the seeds that later produce knowledge and wisdom," Carson said, "then the emotions and the impressions of the senses are the fertile soil in which the seeds must grow."

She was right. She still is. In fact a recent study by the American Institutes for Research shows that kids who participate in outdoor classrooms on average improve their science test scores by a remarkable 27 percent.

As for me, I was lucky. I had a little garden in the backyard, some trees, and a few good teachers. Stuff grew. By the time I was sixteen I could talk about all sorts of things going on outside. I'd take you to the six rows of peas in my mom's tiny vegetable garden, dig down into the dirt with a trowel, and show you a tangle of roots covered in little bumps and nodules. Thanks to my science teacher Mr. Longenecker, I'd be able to tell you about how those nodules are the product of another life-form, a bacteria that secures nitrogen from the soil, and further, that this nitrogen is an absolutely terrific fertilizer. We could talk about how bacteria thrive by feeding on the starches and sugars generated by the pea leaves. And even cooler, that those leaves are feeding aphids, the aphids are feeding the ladybugs, and the ladybugs are feeding the robin that wakes me up every morning of summer singing his heart out from the maple tree.

By age twenty, with two years of college under my belt, I could have told you even more. Had I noticed the ants crawling up and down my mom's tomato stakes, I would've been excited to tell you how trees, especially trees in drier climates, employed ants as bodyguards. And that it worked like this: Scale bugs extract the sugary sap from the leaves for their dinner, doing so without harming the tree, then excrete what they don't use, which leaves plenty of sugar lying around for the ants. So pleased are the ants with all this that they've taken to keeping "herds" of scale bugs, moving them around the tree from place to place like so many sheep. The herds of scale bugs are happy. The ants are happy. The tree is happy too, benefiting from the fact that the ants, which are pretty ferocious, have taken on the job of repelling any interlopers intent on setting up shop and eating the leaves.

Thanks to a few good books and some fine teachers, I could've talked about all that. But mostly it was thanks to the fact that as a small boy I'd been thoroughly drawn in by the colors and shapes of the tulips and geraniums, charmed by the lumbering flight of the bumblebees, enthralled by the humping walk of caterpillars and the squiggles of worms flushed out of the ground after an Indiana rain.

The next time you venture out under the trees, or maybe some dark night when you turn your eyes to a sky shot full of stars, or even when you do nothing more than kneel down in the garden, always right there is the chance to be nudged back into such realms of wonder. It's mostly a matter of getting out of your own way.

First of all, just get quiet. That was Einstein's opening act when he went to the Institute Woods, where he took a breath or two and turned a calm gaze on all the life around him. Admittedly, if you're like me, on a lot of days you may feel like you're caught in a big river of constant obligations and distractions—a fact that can make such calm meditation harder for you than it was for Einstein. The very idea of quiet can be a little off-putting, seeming to open the door to something befuddling—a bit dark and sticky. Sometimes calm and quiet start out with us feeling there's really something else we should be doing. But keep in mind that even mice experience reduced anxiety levels when they're exposed to just fifteen minutes of quiet. Quiet lets us be in deeper play with the world. It allows us to calm down, consigning to the back seat for a few minutes our chattering, anxious, gum-snapping minds. Plain and simple, mystery lives beyond the chatter.

You may not remember, but back when you were a kid you were a master at this sort of deeply engaged focusing. You were by nature an experiential learner—one with no need to shoehorn what you were seeing, hearing, and feeling into boxes somebody else constructed for you. Standing under the flowing arms of a maple, the way you naturally gathered up the world would've allowed you to take in not just branches and leaves and trunk, but the bird and the squirrel and the ants and the sound of the wind and the flutter of sunlight in the leaves. And because at that time of your life the walls between you and the world were thinner, because your culture's habit of splitting off humans from what's around them wasn't yet fully entrenched, the whole scene, in the most wonderful way, also somehow held you— the kid who was doing the looking. Right now you may feel like that kid is long gone. But that's not really possible. Across a lifetime we'll add countless pieces of knowledge and perspective to who we used to be. Yet the ability to be nudged by curiosity closer to the world, to lead first not with intellect but with wonder, is still within your reach. And even better, as an adult you can consciously weave your innate sense of wonder into every aspect of your life—strengthening your sense of contentment, enlivening your relationships. In the end it comes down to engaging your ability to consider the world more broadly and deeply—talents of perception that got obscured by the demands of modern life.

## Putting Your Senses to the Task

Back in the late 1970s, not long after I began my footloose wanderings through the wilds of the Rocky Mountains, I found myself stopping now and then to close my eyes and listen to the wind. In fact I became something of a wind connoisseur. Wandering the Sawtooth Mountains of Idaho or the chiseled peaks of the Tetons, I began to hear the wind as breath—inhaling up the canyons in the morning and then exhaling across the high meadows in midafternoon. What's more, the spill of those winds, fed by the warming and cooling of the day, made different kinds of music depending on what branches and leaves and trunks happened to be living there. There was the hard whispering song of the lodgepole pines, as well as the wavelike hiss of the Douglas fir. There was the stream-like babbling of aspen leaves, which was entirely different from the sharp, driving rhythms of the speckled alder. In the lowlands was the stiff, whipping sound of sagebrush and the contented rustle of wheatgrass. Then, near the top of the world, just below the tundra, I could hear the music of subalpine fir, tuned by the fact that the windward branches were pruned away by harsh weather, leaving only the downwind growth to twist in the air.

By that single, simple act of broadening one of my senses, I began to hear other sounds lifting out of the background: Pinecones chewed off the trees by red squirrels, tumbling down through the branches to land in the needles with tiny, blunt puffs. The gentle squeak and moan of tree trunks rubbing against one another in the distance. The trickle of a water seep. The sweep of raven wings passing overhead.

And so it goes, too, with touch. The cool, moist air by a creek against your skin. The sun's warmth on your eyelids. The furrowed bark of an old oak under your fingertips. The smooth, powdery coating on the trunk of an aspen or paper birch. The cool, dew-covered grass against the bare soles of your feet.

Then there's nature's vast array of scents: The vanilla of ponderosa bark; the lingering sweetness of evening primrose and mock orange; the dry pepper of pine needles and the sharp tang of sage; the sumptuous, intoxicating smell of a meadow after a good rain. Roses. Sweet clover. Dandelion leaves. All best savored with eyes shut, much as we might approach a Sunday morning cup of coffee or a pan of cinnamon rolls fresh from the oven.

There's so much to recommend the sound, smell, and touch of nature, yet most of us tend to be focused on how it *looks*. How else could we swoon over the amber wash of autumn light in a forest, or follow the dance of birds coming home to their evening roosts? How would we be awestruck by deer in effortless flight over fences and fallen trees, or by the procession of shapes forming and dissipating in the summer clouds? Still, when it comes to the mysterious, relying on vision alone can be a bit of a liability. A bridle on the brain.

As it happens, that particular bridle was first yoked to our powers of attention by the early Greeks.

Imagine walking through ancient Athens some fine spring day. Long, well-tended strips of garden border the cobbled pathway, threaded with apple blossoms and marjoram and thyme. Just ahead, around a slight curve, a set of stone steps leads down to a small theater where the much-admired scholar Anaxagoras is sitting with a dozen eager young students. As often happens, one of those students—curious in that restless, urgent way of youth—cuts right to the chase:

"What is the reason for a human life?" he asks Anaxagoras.

Anaxagoras doesn't miss a beat.

"To look," he says. "To look at the sky, the stars, the moon, and the sun."

The scholars of ancient Greece did an awful lot of looking. And by means of their patient, dedicated eyeballing they managed to puzzle out all kinds of things: divining an explanation for lunar eclipses, speculating about meteors and lightning and rainbows, watching how water moved and harnessing it to drive everything from mills to pipe organs. They took hard, deep looks at things with the intention of discovering verifiable reality. And that proved the foundation for enormous accomplishments. Our own science still revels in it to this day. A century after Anaxagoras sat on those stone steps teaching his students, Aristotle would declare that the most perfect existence for a human was the theoretical life—from the Greek word *theoria,* to *look.* And specifically to look in a way that was attentive but at the same time always independent. Clearly removed from what was being looked at.

It was a specific kind of vision. And now it's ours. Deeply rooted in the belief that there's an observer separate from the observed. The so-called objective gaze. A means of disconnection. The bridle.

No need to discount what such a learning approach can yield. Still, it's very good news that today's science is jettisoning the belief that this kind of seeing represents all we can ever know. While it's taken us a very long time to admit, objectivity never was the whole story.

Think of a three-year-old girl finding herself one sunny

morning at the foot of a big pine tree—one she's never seen before. The ancient Greeks, and those of us who followed in their footsteps, would say what she learns about the tree in that moment comes from straightforward, outer-focused observation. This tree is taller than the one over there but shorter than the one in her front yard.

But now we're starting to understand something rather different is going on. Instead of merely observing the world, she's likely experiencing the uprightness and height of the tree in front of her by subconsciously *feeling* it as it relates to her own position and height in the world. In other words she's having not just a cognitive brain experience, but one driven by her whole body.

Her connection with the tree gets sent to her brain not as a photograph that can be compared to other tree photographs, which is what the objective gaze expects, but as multisensory packets of information. Twenty years later, given the right trigger—maybe the smell of sap, or even craning her neck to look at the upper branches of another tree—those packets from that pine tree long ago may be instantly animated, enriching the young woman's moment with a rather mysterious blend of comfort and complexity.

Rather than our thoughts being disconnected from our bodies, as we've assumed from Greek times, they may at times be *driven* by our bodies. That suggests an interesting idea. By putting your body in nature, and once there engaging the full range of your senses, you may well be assembling powerful packets of symbols by which you can lay down a very different, very intriguing new way of experiencing the world and its mystery in the days and years to come.

Closing your eyes every so often and employing touch, hearing, smell, or even taste breaks down your visual bias. Scent and touch and sound are less bound up in the way vision is. When we close our eyes and nudge our noses against a wild rose bloom, after all, there's really not any notion of there being a smeller and a smelled. Instead we end up engaging in a way that allows the enchanted sensation of mutual embrace.

## Thought Streams

It's autumn in the woods. You know the feeling. Summer gone, winter not yet knocking on the door. The world exhaling. Quiet contentment with a dash of melancholy. Maybe the moment brings on a happy memory of being bundled up in a sweatshirt playing touch football, or diving with abandon into piles of leaves. Or if you're a nerd like me maybe it's more a matter of getting caught up in what's actually causing that autumn smell—how the leaf litter is being gently ingested by fungi and bacteria, or how the cold air has quieted many of the competing smells normally in the landscape, thereby bringing to the foreground this wistful scent of decay.

Any of that can be fun, interesting, satisfying. But the

next time you're outside and those thoughts show up, see if you can savor them for a minute and then let them go. Let them float by like so much dandelion fluff. See if you can perceive the world around you from a higher perch—one that affords a view of life beyond your stream of thoughts.

## False Choices

A few years back, in a small town in Montana, I was in a café sharing a cup of coffee with a ranching couple in their late forties. The man was a little on the cranky side that day, annoyed at what had become a flood of newcomers pouring into the state.

"They fall in love with the wide open," he grumbled. "But they never stop to think that it's wide open because of farming and ranching. They move in and then end up complaining about the plow dust and the bellerin' cattle. They're all crazy."

His wife looked at him and shook her head. "Some do that," she said. "But not everyone. Saying all of them is just lazy thinking."

"It's not lazy," her husband replied. "It's efficient."

She's right. But in a way, so is he. There's a lot of brain research right now to suggest that most of us are extremely prone to exactly the kind of low-energy thinking exhibited by that rancher. During conversations we often employ a kind of shorthand, a black-and-white way of describing the world that takes very little effort. And we do it with even more gusto when we think we're talking to someone who sees the world like we do. We categorize entire groups of things, including people, as having one prominent set of qualities when in fact any group is nuanced and complex. When in fact they're nothing short of a bona fide wilderness.

We feed these "categorical imperatives," as they're sometimes called, with deeply entrenched "binary" thinking: liberal or conservative, smart or ignorant, quick-witted or slow, straight or gay, black or white, good or bad, us or them.

That habit, it turns out, comes from a part of our brains known as the frontal lobes. Frontal lobes are definitely handy. They're what help you identify a problem at work and then brainstorm with your co-workers how to go about fixing it. What's more, your frontal lobes let you show back up at the office the next day with a sound memory of what everyone said in yesterday's meeting, and then actually get down to the work of applying solutions. Then that evening, on your way home from work, if you see cars backed up to Hampton Avenue, past experience may tell you that the northbound interstate is likely choked with traffic. Using your frontal lobes, you decide instead to go by way of Martin Luther King Boulevard. By employing frontal lobes, we can segment the world, holding in relative isolation those things that need immediate attention.

Even animals do it. Mervin the cat, having suffered even once the wrath of Hank the bulldog next door, will file that experience away and run off like a shot at the mere sight of him. By categorizing Hank as dangerous, Mervin keeps from getting thrashed. Likewise, wolves that have been exposed to hunters and trappers, witnessing firsthand what human hunting does to their packs, will quickly learn to go into extreme stealth mode whenever people are around. By contrast, if your golden retriever—a distant relative to those wolves—gets treats from your sister whenever she comes by your house, rest assured he'll be at her side as soon she walks through the door. Your dog categorizes your sister—and possibly anyone with her—as a fun human packing treats. And that works out fine. Categorizing comes from a place of perceived certainty. It attaches to things that you, or your cat or your dog or those wolves, don't have to invest time and energy thinking about.

But as useful as categorizing is, that usefulness can fall apart fast when we apply it to things about which we can't be certain at all. Like any part of life that contains mystery. In categorical thinking the boundaries are drawn. But mystery is open-ended, messy, full of promise, and lacking in certainty.

Making too much of your thinking categorical, walling off any possibility that life has a kind of unknowable wildness to it, can lead to sadness and discontent. Cambridge University psychologist John Teasdale found that therapy patients with an "absolutist, dichotomous thinking style" (what I'm calling categorical thinking) were at significant risk of depression. Likewise, neuroscientists at the

University of Reading in the United Kingdom discovered that the use of absolutist words in language—like "you always" or "every time" or "never"—may predict mental distress. In their study of sixty-four hundred people in online mental health chat groups focused on personal depression and anxiety, absolutist language was 50 percent higher than in the population at large. In suicide-focused groups, it was 80 percent higher.

Categorical thinking and binaries make life brittle and less interesting. And as Harvard chemistry and physics professor Eric Heller reminds us, "Be careful how you interpret the world. Because it *is* like that."

There's just no better place to go for a vacation from binary and categorical thinking than nature. Even if our habitual minds lead us into the woods with thoughts that deer and hawks and strawberries are good, while flies and mosquitoes are bad, any real inquiry, while not making us any fonder of the bugs, would yield at least a slightly expanded sense of nuance. Flies are the primary pollinators for everything from orchids to trilliums, and what's more, as decomposers they're absolute masters at preparing dead things to become live things again. Flies eat so many aphids and moth caterpillars that it's probably safe to say they're largely responsible for the success of the world's agricultural crops. Mosquitoes, on the other hand, feed everything from fish to lizards, birds to salamanders. Meanwhile the kin of those pesky midges (so-called ceratopogonids) are a major pollinator of cacao. Getting rid of them could mean—*gasp!*—the end of chocolate.

Hummingbirds good, starlings bad. Chipmunks adorable, rats and mice—at least beyond the cartoon variety—a travesty. None of it makes sense when you start looking closer at the workings of the natural world. No wonder people in cultures far older than our own have long found such divisions—good versus bad, cute friend versus ugly pest—the sign of badly broken hearts and minds.

You don't have to fall in love with flies. But it might be worth entertaining the notion that flies (and dandelions and wolves and skunks and crabgrass) were created by the same proteins, the same basic planetary reach for life, the same deep and inscrutable web, that created you and me. Go into the woods with your insect repellent on, swat the flies and mosquitoes as you will, but see if you can resist getting caught up in wishing the world was some other way. At the end of the day, all of us just are.

Nature can bust binaries in other ways too. Think for a minute about the fact that the vast majority of the flowers and plants that grace the Earth are hermaphrodites. The grasses in our yards, the lilies, roses, squash, corn, and cucumbers in our gardens, possess both male and female reproductive systems. Meanwhile there are an amazing number of fish, jellyfish, limpets, and other shellfish that can change from male to female on demand, and vice versa, depending on what their communities need at any given time. That one binary alone, then—that everything must be male or female—is an illusion. And like so many of our illusions, it's another layer of bricks in the wall that keeps us disconnected from the ways of the world.

Wandering the wilds, I've grown fond of seeing nature through the ancient Chinese symbol for yin and yang: two tadpole-shaped figures embracing each other head to tail, one black and one white, one masculine and one feminine, the pair bound within a ring symbolizing the circular reality of life. Yin and yang are not opposing but complementary binaries; taken together they yield more than the sum of the parts. Then there's that dot of white in the black figure and that dot of black in the white one, looking a little like eyes, reflecting the fact that each contains a piece of the other. Finally, at the exact center of the yin-yang circle, along the curved line formed where the images come together in their embrace, is the still point: that place of peaceful mind, body, and spirit that, appropriately enough, is said by the ancients to be sustained by the quiet contemplation of the mysteries of nature.

## A Matter of Time

If you live in the Northern Hemisphere like I do, it would seem awfully weird to run across of a group of wolves hanging out on the elk's winter range in the middle of July, when the elk—the wolves' primary source of food—are miles away, high up in the mountain pastures. Likewise, you wouldn't go to the blackberry patch in April, trek off in the middle of winter to net spawning salmon, or try to coax bees to the work of making honey when there aren't any flowers in bloom.

Yet we're living out our days in a world of instant messages and next-day shipping, of more movies, TV shows, podcasts, and music on demand than we could ever hope to watch or listen to. We can pull out our smartphones and turn up the heat in the house when we're still miles from the driveway, pause in the middle of putting away the groceries to conjure up a date for later that night with the swipe of a finger. A medical specialist two thousand miles away can videoconference with our doctor about our X-rays while we read the latest issue of O Magazine in the waiting room. Plain and simple, the cleverness of our technologies has led to our getting a lot of what we want when we want it.

Reclaiming the delights of the natural world—to make what's delectably mysterious "out there" available to us "in here"—means taking stock of our relationship with time. As Iowa State University geology professor Cinzia Cervato reminds us, the development of clock time, which really only became a thing in the Middle Ages, changed profoundly our experience of life "by making mathematical units seem more real than lived reality." And that has a lot to do with why mystery, at least for many of us, can be awfully hard to rouse.

Not that nature can't be fast. Earthquakes. Volcanic eruptions. Lightning. Wildfires. Flash floods. But the natural world rests on an inescapable foundation of long-term expressions. A coast redwood will reach for the sky across a thousand years, going from a tiny sapling nursing on a fallen log to a mature tree weighing an astonishing twelve million pounds. Rivers take millennia to carve out new routes to the sea. Mountains are both rising and crumbling at the same time, growing and shrinking by the inch over millions of years. Hiking across the long sweeps of alpine tundra on the northeast edges of Yellowstone, I'm routinely humbled by the fact that it takes more than a thousand years to make just the top inch of soil under my boots.

Then there are the yawning stretches of time needed to navigate the reaches of space. When we gaze into the night sky, because of how long it takes the light from those stars to reach our eyes, the scene we're swooning over is actually one from decades or even centuries in the past. Lingering in nature helps us make peace with this more fundamental pace of emergence, letting us put away clock time for a little while and roll instead with the rhythm of *physical* time.

It's amazing to think it's been only about 250 years since even the scientists among us really embraced physical time—which refers to the procession of events since the planet swelled into being. Prior to that, most people were content with the idea that the Earth was about six thousand years old, the beginning of life on the planet having unfolded across a series of catastrophic events launched by

God, each marking the birth of a new age. To be more precise, back in the 1600s, using calculations taken from the Bible, Bishop James Ussher of Ireland pegged the first day of creation as Sunday, October 23, 4004 BC.

But in the late 1700s there came onto the scene a dogged amateur geologist from Scotland named James Hutton. And Hutton thought otherwise. After years spent wandering and studying the rocks of his home landscape, he became convinced that stone eroded slowly, with the resulting sediments eventually buried and then—again, very slowly— turned once more into rocks by pressure and heat. The picture of Earth he painted was more that of a constant loop of creation with, as Hutton himself described it, "no vestige of a beginning, no prospect of an end." We now know that he was pretty much right on the mark. But back then, and for a good long while afterward, his ideas were shocking.

Some called his theories an outrage, accusing him of being a heretic. An atheist. For others, though, his notions were a delight. Mathematician John Playfair, one of Hutton's contemporaries, described himself as "giddy" at having found the opportunity to look so far back into the abyss of time. Arguably it's thanks to the groundbreaking discoveries Hutton made that you and I are better able to perch our imaginations on the edge of an almost unimaginable expanse of creation.

To cast our thoughts into the abyss, to knowingly walk on a planet that's been renewing itself time and again through massive geologic processes stretching across millions and millions of years, seems an especially fine place to first obliterate and then profoundly broaden our perceptions. Letting our imaginations spend time in frames too big to grasp offers us the chance to access ever more of the world's wonders.

From there we may well open again to notions of endless spoolings of rain and then the return of sun. To the countless number of mornings that birds have come to their branches in the light of dawn and launched into song. To the four billion years of the moon swelling and receding, pulling and then releasing the ocean tides. These ageless circles are a powerful portal to all things mysterious. What's more, in contemplating them you may find yourself more frequently observing your own smaller rhythms—the rising and falling of energy across the day, the week, the year; the hunger you feel for dinner and then the fullness; the grace of your body moving and then resting again; the pull of relationship and the comfort of solitude; the slender pulsing of blood in your neck and wrists; the rising stature of your children as they grow; the whisper of breath coming in and going out. And if you're lucky, the deepening of gratitude for all things beautiful as youth turns into middle life, and midlife becomes old age.

I was in my late twenties. My mother was bedridden, dying of cancer. For a good month or more she'd been too weak to even hold her head up. But one morning, as I sat at the foot of her bed, she suddenly perked up a little, then told me she wanted to go outside. So I carefully gathered her up in my arms and carried her through the front door and out into the yard. Around we went for what must have been twenty minutes—first so she could smell the flowers on her lilac bushes, then so she could look through the woods above the bird feeder for the flash of a certain cardinal's wings. And finally, so she could run through her fingers the supple young leaves of the maples and dogwoods.

We said very little in those moments. But by some extraordinary grace she managed on that day to take in the mystery floating through that little yard and use it to light the dark place closing in around her. Incredibly, that afternoon the pain she'd worn for so long began draining from her face, replaced by a look of serenity I'd never seen in her. The next morning she told us to stop all the painkillers, and this despite having been on massive doses of Percocet and morphine for several months. A few days later, in the still hours of the night, she drifted away.

I would later recount that small and tender journey, my mother in my arms, to a well-known Jungian analyst. When I finished, she told me about one of Jung's observations— how he noticed that during religious sacraments people sometimes experienced a deep sense of mystery. And what's more, that such mystical experiences were often a powerful boon to their mental health. Wading into that kind of mystery stirred the unconscious, Jung said. And thus stirred, mystery became for many a powerful force for healing.

"Now, think for a minute," the analyst proposed. "That little yard of your mother's, filled with some of what humans first used as religious sacraments. Trees. Flowers. Birds." She asked me to try to imagine what it might be like to be really immersed in those threads of nature. To envision sinking in, feeling welcomed into the very flow of living and dying—to know, even for just a few seconds, the sensation of actually *participating* in the mystery that binds us all.

It's a bit easier to imagine that in times of extraordinary trauma, like the death of a loved one, we'd stand a chance of being broken open by the pain of our loss, suddenly finding ourselves at the edge of something inexplicable. But the truth is we can also befriend the mysteries of the world in less tumultuous times, learning to ride its currents into a place beyond our daily dose of doubt and anxiety and frustration.

Seeking mystery means sailing now and then across some deep waters, free of the need to know what weather lies ahead, or even what fish are swimming under the boat. For some one hundred thousand years, we as a species have been walking around in more or less our current form. And across all that time nature has been overflowing with the mysterious. Granted, much that's precious in the natural world has been preserved because of highly rational arguments that prioritized what we could least afford to lose. But if our goal is to figure out how to reawaken our instincts of kindness toward the world writ large, to create in ourselves a more powerful perception of and connection to place, to savor the deep joys of belonging—that calls for something else entirely.

# Life on Earth Thrives Thanks to a Vast Garden of Connections

The man killed the bird, and with the bird, the song, and with the song, himself.

—Pygmy legend

The story goes something like this. Nearly four hundred years ago, a bitterly cold night in the Netherlands. The air is stung through with hard wind and sheets of snow. Yet the weather hasn't stopped throngs of students bundled in waistcoats from filling the stately Utrecht University auditorium, eager to hear a lecture by a man widely touted as the most intelligent human on the planet. His name is René Descartes. And while for years he's been celebrated as a philosopher and mathematician, tonight people have come because of his growing reputation as a "cultivator of science." An intellectual whose tireless rational inquiries have allowed him to divine God's master plan for the universe.

Beyond the flickering light of the whale-oil lamps ringing the stage, a professor in a black frock struts out to quiet
the crowd, announcing in a booming voice that the brilliant
René Descartes will soon appear and transform the way
they see the world. And what's more, the professor says,
some in the audience might be astonished by the power of
what they're about to hear. The professor squares his shoulders and exits. Five minutes pass. Then Descartes strolls,
regal, into view. Following him onto the stage is a sluggish
yellow dog. On Descartes's command the dog lies down in
front of the podium, his head between his front paws, staring past the oil lamps into the darkness.

Halfway through the lecture, deep into describing the
workings of the universe as one might explain the gears and
springs of a clock, Descartes pauses. He stretches out an
arm, the index finger of his hand pointing to the rafters as
if to hallow his proclamation that humans are the only species capable of deciphering the genius of the Creator. Humans are exceptional, he says. We are chosen. And not only
is every other creature of inferior intelligence, but none
has emotions, or thoughts, or even the capacity for physical
sensations.

It's at that moment when Descartes walks around to the
front of the podium and kicks his dog. Kicks it hard.

The dog squeals.

Against an audible gasp from somewhere in the audience
he holds the palm of his hand. He smiles like a patient father.

"This I can tell you with certainty," he might have said.

"Modern study has affirmed that no creature on Earth save humans has any ability to feel pain or fear. What you are seeing is called an automatic physical response. Because of its limited nervous system, the dog experiences no discomfort whatsoever."

Then, to emphasize his point, Descartes kicks the dog again.

René Descartes is considered one of the most brilliant men in Western history—described often, and rightly so, as the father of modern science. And while we may be horrified at the thought of a guy going around kicking his dog in front of packed auditoriums, the truth is we still embrace much of the fundamental thinking behind his antics. Plenty of us still coddle a wholesale belief in human superiority, as if living beings should be ranked according to their worth, which does much to alienate us from the rest of life. For a long time now we've considered ourselves exceptional beings at the center of a universe made up of identifiable parts, the whole lot of them there for the taking. A nature that seems more mechanism than organism. Even today, what Isaac Newton said 350 years ago, calling God "a highly skilled mechanic," is still to many of us about right.

The worldview laid down by the likes of Descartes and Newton—clever, precise, and in hindsight astonishingly overfocused and overconfident—didn't come out of nowhere. Such brilliance, and it really was brilliance, sprouted some fifteen hundred years earlier, tended by the minds of the ancient Greeks. It was the Greeks, after all, who hatched the idea that the universe was bound by a set of fixed "ultimate truths." And further, that if you hoped to discover those fixed truths you'd need to hunt for them with a mind fiercely trained in rational, objective thinking.

For the Greeks, any question about how nature worked could only ever be satisfied with a single answer. And if what you wanted to know about the world couldn't be so answered, if it didn't map tightly onto one of those single truths, then, well, the problem was with your question. Getting to the ultimate truth of nature meant casting your focus "out there," using your mind to break things down into categories and component parts. And that approach, which Descartes and others of his day supercharged with some amazing mathematics, moved quickly out of the realm of theory, of intellectual story, to become reality itself. If you couldn't establish the clear physical boundaries of a thing—hold it outside yourself, fixed in time and space— then it didn't exist.

In order to discover the "real" properties of nature, Galileo said in the 1600s, sounding himself a bit like an ancient Greek, one had to see the natural world as if humans didn't exist. Understanding the world had absolutely nothing to do with our personal experience of it. Wisdom was to be gained by focusing on the objects that existed apart from us, the qualities of which were best expressed in terms of mathematical relationships. The universe was "real" only to the extent that it could be objectively studied, measured, and ultimately predicted. And in one sense he wasn't wrong. It was just that sort of approach that led to Galileo's famous *Starry Message,* revealing discoveries he'd made with his handcrafted telescope of mountains on the moon, hundreds of stars previously unseen by the naked eye, even the moons of Jupiter. He expanded human perception and in the process seeded no end of wonder.

Yet getting hyperfocused on discrete objects, coming to believe that kind of structuring held the key to the whole story of everything, dimmed our ability to perceive dynamic relationships. And that would cause us—is causing us—to miss an extraordinarily important fundamental truth about nature, which is that no life exists except through vast, dynamic webs of connection.

My own grade school and high school science classes taught me to see the world the way early science did. Memorizing as many of the defining details as we could about a single thing, from a plant cell to a frog's heart, and then moving on to something else. My college classes, thankfully, offered something more cutting-edge, more centered on what was then a fast-rising wave of interest in relationships. In the *ecology* of life. I took that view along whenever I went into the woods, holding it up to what by then had become for me an exquisite, even giddy obsession with nature.

About ten blocks from my parents' tiny house in South

Bend, Indiana, across Mishawaka Avenue and past the deep brushed green of the Nuner Elementary School football field, there was a patch of mature forest not far from the banks of the roiling, chocolate-colored St. Joseph River. Lily of the valley and wild licorice were tossed across the ground, just like when I was a kid, and overhead the fluty songs of orioles, finches, and cardinals dripped from the branches. Most striking of all was a copse of old oak and maple trees, some bigger than you and I together could put our arms around.

During my senior year of college I went down there fairly often, mostly just to be with the trees. If I happened to come on the heels of one of my environmental science classes, I'd be prone to giving them a shout-out for the carbon they stored, which at the time was being talked about nonstop in the context of something called global warming. If I'd been reading Darwin, on the other hand, I might have tried to wrap my head around the fact that those old trees on the St. Joseph River had as a species been fine-tuning their lives for some twenty-five million years, coming online in the era of saber-toothed tigers and forty-two-foot sharks.

But sometimes I was just happy to consider my old tree friends more whimsically—through the lens of how our ancestors saw them, full of daydreams and solace and myth. The giant bodhi tree where Buddha snapped to enlightenment, and the low-hanging branches of a terebinth, where Abraham was said to have entertained the angels. Zeus at the oracle whispering advice to mortals through colossal oaks; the kings of Egypt laid out in their soundless pyramids, wrapped in coffins made of sycamore, said to be a tree of nourishment for the souls of the dead. Glooskap the Giant, from Iroquois myth, roaming the woods of Maine shooting arrows and splitting open the trunks of trees to release the first men and women into the world. And American landscape painters who for a time filled their canvases with mammoth elms—our national tree—a symbol, they said, of the promise of freedom, of democracy in the New World. All of us, really, joined together by our stories of the trees.

As I grew older, the woods seemed richer still. In part because a rapidly expanding scientific understanding has given them a shimmer as bright and compelling as any legend or myth. We now know that conifer trees have larger, more complex genomes than almost any other life-form on Earth. That big spruce tree you may walk past now and then in the town square has, as a species, been quietly going about its business for several hundred million years, which has helped produce genetic material about seven times bigger than your own. And we know, too, that individual trees can have extraordinary longevity. Not just the bristlecone pines of California, some four or even five thousand years old, which means they first sprouted around the time Stonehenge was rising. But also the root system of the 106-acre Pando aspen grove in Utah—the largest known life-form on the planet—now thought to be an astonishing eighty thousand years old, which places its starting point at roughly the time humans were leaving Africa on their epic journey to settle the world.

But more than any of that, trees are exceptional portals into this particular master lesson of nature, offering us eye-popping glimpses of essential interdependence. Right this minute in that oak grove near the St. Joseph River, information about invading bugs is being broadcast between trees by means of airborne hormones—a trick that helps coordinate making defense compounds in the stems and leaves. Some of the trees, struggling against chomping caterpillars, are releasing pheromones as a signal to call in nearby wasps. The wasps then buzz over and lay eggs; their offspring, in turn, end up dining on the invading caterpillars and thus save the trees. Apple trees under similar attack get even faster relief, releasing chemicals that signal caterpillar-eating songbirds. This sort of green talk, expressed through a language of chemical exchange, is going on all over the planet, wafting through the air from the scrub birch of the Arctic to the jungles of South America; from the mountains of China to the golf courses of Tennessee. And what's more, such talk holds hundreds of messages we've barely begun to decode.

Then there's the matter of a major miracle going on in the soil. The trees of the woods sport vast tangles of root systems. Those roots are in turn connected by an entirely different life-form known as mycorrhizal fungi. Forests and fungi got together millions of years ago, and over all that time they've managed to fashion a highly evolved relationship that's hugely beneficial to both. Fungi actually preceded trees by about a hundred million years; some even looked a bit like tree trunks, with the aboveground parts growing several feet wide and twenty-five feet tall. The kind growing under the woods, though, employ tubed tendrils about the thickness of spider-web silk to weave themselves into the root structures of the trees. By doing so they gain access to the sugars the trees produce through photosynthesis. The trees, meanwhile, use this big web of fungi to gain access to essential nutrients like nitrogen and phosphorus that otherwise wouldn't be available to them.

When a fungal web first attaches itself to one of the root systems, it kicks off a defense response. This may be the equivalent of kick-starting a tree's immune system—sort of like when we get a flu shot—leaving the tree much more resistant to disease. But beyond that, the underground fungal webs are allowing individual trees to actually communicate. To quite literally help one another. Turning the forest into a neighborhood, in the very best sense of the word.

The last time I visited that grove I saw several little seedlings in deep shade, struggling a bit. They were having trouble gleaning enough sunlight to make available the carbon they needed to grow big and tall. Many of the other, bigger trees would've been "aware" of this particular problem because they were in touch with the saplings through the fungi. Some would respond to the problem by sending carbon and other nutrients, molecule by molecule, through the network, thereby passing nourishment from those who had enough to the little ones who didn't.

One tree at the edge of this grove was showing early signs of being attacked by blight. Almost certainly that tree was giving a heads-up to the rest of the woodland community by sending chemical messages through the fungal communication network; those messages would stimulate in the rest of the trees a defense response, thus greatly reducing the blight's damage across the woods.

One especially large oak I came to know in the St. Joseph riverside stand, an old matriarch almost twelve feet around, likely played a role best described as that of a "grandmother" tree—serving dozens of other trees, including quite a few of other species, by routinely releasing nutrients for the young and vulnerable. When it comes time for this grandmother tree to pass away—and from what I saw of the wounds in her trunk, she probably doesn't have a whole lot longer—as she dies she'll use the network to send her share of resources to other residents of the neighborhood.

With every passing year, the forest teaches us more. We've recently come to understand that a by-product of these fungi going about their lives is that they produce a wide array of antiviral compounds. One summer on a walk in southwest Maine, near King and Bartlett Lake, I happened across a large gray birch tree. Looking closely I could see slashes on the trunk—likely from a neighboring tree that was blown down, scraping the bark of the birch when it toppled. The resin that leaked out of those cuts proved a great habitat for polypore mushrooms, which had set up shop on the scratch marks. Local bees had found this wound in the tree, too, and had put themselves to feeding on the sugars in the sap. In the process they were ingesting antiviral compounds from the mushrooms, which in turn would help them stave off threats like mite infestations back at the hive. Of our current efforts to combat the catastrophic decline of bees around the world—due in part to two deadly viruses carried by varroa mites—one especially effective treatment is using an extract from mushrooms. Indeed, applying that extract to the so-called Lake Sinai virus diminished it an amazing forty-five-thousand-fold.

So the takeaway here is that trees and plants aren't simply competing with one another, as I came to assume during my junior high science classes. Instead, over millions of years vegetation has built vast collaborative networks to allow the system as a whole to thrive. And that system includes us.

When we walk along a woodland path, the trees and even some of the smaller plants growing at our feet are giving off invisible airborne antimicrobial compounds called phytoncides. We inhale them with every breath. Once inside our lungs, those phytoncides are passed along to the cells of specific nerves and arteries, resulting in a reduced heart rate and lower blood pressure. They're also being taken up by our lymph glands, where in short order they can be used to boost a flagging immune system. Still others are being channeled to the hypothalamus, where they'll help support the more efficient functioning of our vital organs. So undeniably valuable are these and other benefits of trees to humans, says University of Illinois biologist Dr. Frances Ming Kuo, that "the less green a person's surroundings, the greater their risk of disease and death."

After hundreds of years we're finally beginning to shed the illusion that we have the natural world in our hands, slowly but surely replacing it with the far more accurate, more comforting truth that in so many ways, the world has our backs.

Though he may have lacked scientific knowledge of the forest, poet Walt Whitman, struggling after a severe stroke that paralyzed half his body some 150 years ago, spoke aloud to a patch of forest outside Camden, New Jersey:

"Hast Thou . . . medicine for a case like mine? . . . Dost Thou subtly mystically now drip [that cure] through the air invisibly upon me?"

If we can put on hold now and then our habit of focusing on single life-forms—a squirrel, a tree, a white-tailed deer—in favor of seeing the surrounding context of those things, what we really end up doing is building mental muscle for reintegrating our lives with the world at large.

Years ago I got a wonderful opportunity to experience this firsthand. I'd been asked by *National Geographic* to create a written portrait of the most remote place left in the Lower 48—defined as that location farthest away from any roads. As it happened, this place was located in the extreme southeast corner of Yellowstone, separated from my home by a long, massive run of mountains.

So one day in early summer I stepped out onto my front porch, shouldered a pack, and began a 125-mile walk across the mountains. I reached that most remote place ten days later, walking the final miles through a wild and beautiful valley known as the Thorofare, along the upper reaches of the Yellowstone River. It was a thrilling journey—a feast of wolf sightings and grizzly tracks and soaring views from the tundra in every direction. I stayed in that high country for nearly three months, most of it in a tiny Forest Service cabin, finally winding my way back home in autumn, just as the bugles of bull elk in rut started ringing down the mountain valleys.

In the early days of that adventure, I mostly drifted between delights: from waterfall to fishing hole to soaring mountain view. A beautiful bear track was a beautiful bear track. An elk lunging through the lodgepole was exactly that, as was a bald eagle fishing the slack waters of the Yellowstone.

But the longer I was there, the more it became clear that patches of trees and runs of sagebrush and even creatures themselves were tethered to entire communities, their lives unfolding out of the circumstances of countless other characters. When I saw elk moving fast late in the day, if I was patient and hid myself, as often as not I'd see either a bear or a pack of wolves trailing along in their wake. Well equipped for that exact circumstance, the elk had picked up the predator's scent and started their escape.

Looking a little more deeply, I could see the elk were standing where they were because of a fabulous weave of new growth along a spring creek, which was running strong because of a good snowpack the year before. The same snowpack that had kept the bear in his den a little longer than in some years, depleting more of the fat he'd used to sleep away the winter. Opening my eyes even wider, I saw that if it was wolves following the elk herd, there would usually be ravens drifting along overhead, those birds understanding well that should the wolf pack bring down prey, they'd be among the first in line to pilfer leftovers.

Then there was the weather to consider. The moisture in the air affected the degree to which the wolves and elk could catch each other's scent. A sudden heat wave rolling into the mountains in early August put the bears and wolves into their day beds, not moving again until the cool of twilight, which kept predator and prey apart.

As summer wound on, I became aware of the setting sun

drifting slowly southward, touching down on the western horizon in a soft hopscotch from dimple to notch to peak—pushing me to light the lanterns slightly earlier at the end of every day. I saw how calm evenings brought the mosquitoes and gnats that gave rise to bevies of cartwheeling swallows and nighthawks on their daily quest for bugs. In August I tracked the level of the river quietly slipping down the banks as the last of the snowpack melted off the mountains' slopes. Snows that had watered the grasses that fed the elk, and at the same time kept the creeks flowing, giving life to cutthroat trout.

Trout that in turn had fed the bears in June and the otters all summer long, and also coaxed fishermen to venture some twenty-six miles by packhorse to try their luck with fly rods. I imagined that snowmelt running on for nearly nine hundred miles to join the Missouri River, and then the Mississippi River and on to New Orleans, where it would float the ships from South America that brought the coffee I was pouring into a tin cup, warming my hands as those mountain mornings began to chill.

Science has done a good job investigating the biological strands of that wildly enchanted country. Hydrologists and ichthyologists, wildlife biologists and botanists and entomologists, have all been there. And to be honest, I love lapping up the stories of their research. But what being out there day after day in the Thorofare did for me went beyond the specific facts of water or bears or wolves or rivers or night-hawks or sky.

"Focus!" Descartes might have yelled at me from those meadows in Yellowstone. "Isolate. Disconnect the object from everything else. Form one question. Find one the an-swer. Put it in a box."

But the Thorofare of Yellowstone, like all of Earth, is a theater with no audience, where every single thing, includ-ing humans, is among the countless actors in the play. In that one valley alone was a web of life so clearly full and dynamic that any attempt to zero in would only capture moments that had already floated away. I could know some-thing of bears, all right, but I'd never really know what there was to know about *that* bear. And clearly, I couldn't be merely an observer, as Descartes and the Greeks before him would've encouraged.

Because I, too, was one of the actors. For one thing, I was an unwitting courier for hundreds of foxtail barley seeds hitchhiking rides on my socks for the chance to sprout and grow in places they'd never been before. And beyond that, the mere sight of me out walking a trail brought curi-ous magpies in for a closer look; their arrival, in turn, caused songbirds to cling fast to their nests, fearing the magpie might scavenge their eggs. At the same time, voles and mice would've scurried off to keep from ending up as a magpie's lunch. Meanwhile the scent I gave off no doubt altered the plans of many a wolf, and probably a few bears too, sending them dashing off in the opposite direction from where it was they first meant to go.

True, the wider view of life I glimpsed in the Thorofare is with increasing frequency being fortified within the walls of a thousand contemporary biology and even physics labs around the world. But it's a view that also holds loud echoes of ideas laid down long before Newton and Descartes—anchored thousands of years ago, in fact, across countless indigenous cultures and spiritual traditions. And those traditions are still very much with us today.

Over in the Plum Village Monastery in southern France, for example, celebrated Zen teacher Thich Nhat Hanh stands in front of a classroom of students, holding in his hands a single sheet of blank white paper.

The poets in the group, he'd very likely say, will be able to see clearly that this paper contains a cloud.

"Without a cloud there would be no rain. Without rain the trees cannot grow. And without the trees we cannot make paper."

He lets that settle in. Then he tells them that if they look more deeply still, they'll see sunshine in that paper too. No sunshine, no forest. And without a forest, no paper.

And then he goes further still.

"If we continue to look we can see the logger who cut the tree and brought it to the mill to be transformed into paper." Look some more, he tells them, and you can see the wheat of the logger's daily bread. Given that this logger can't exist without his daily bread, we might say the wheat that became his bread is also in this sheet of paper. And you and me, we're in that paper too, Thich Nhat Hanh continues, because paper is part of our perception.

"Your mind is in here also, so we can say that everything is here within this sheet of paper. You cannot point out one thing that is not in here—time, space, Earth, the rain, the minerals in the soil, the sunshine, the cloud, the river, the heat. Everything coexists with this piece of paper."

After three months neck-deep in the Thorofare, I knew that he was right. And also, that everything coexists, too, with the lodgepole pine. The wolf. The raven. The elk. The bear. My time in that outback offered me a kind of course correction. I started to put down my unquestioned faith that everything knowable was available with the application of rational, objective science. I also started wondering what the world might feel like if we held the kind of perspective that still shines through many indigenous languages, where life is described more in terms of verbs than of nouns. So rather than telling someone you saw a deer, it's more that you experienced the various forces of life coming together to express themselves in a creature that's "deering." Likewise, a man could be said to be "manning."

What if we could free ourselves from the confines of certainty—to learn to dance with the fact that reality zigs and zags across shifting ground? Increasingly, I'm seeing the great relief that can come from letting go of our insistence that the world be fixed in bound forms, each fated to act within a narrow range of predictability. Instead, we can learn to awaken each day much as some of the best scientists do, more excited with the new questions the morning brings than with the answers found yesterday afternoon. Delighting in the fact that the learning never ends.

There's a kind of sweet comfort that comes from realizing that life on Earth is less a wrestling match than a science fair. Yes, there's plenty of competition in nature. But as is true with humans, a fair amount of that competition consists of coming up with better strategies for taking advantage of changing conditions. As Darwin himself said, "those communities which [include] the greatest number of the most sympathetic members . . . flourish best."

Yet when you grow up as so many of us did, in a world that still has one foot in the Greek habit of isolating and ranking things, it's easy to be misled on exactly what it is that drives human lives. In the United States we've been remarkably fond of leaning on slogans like "survival of the fittest"—which in itself is a bit of a misnomer, since originally that phrase referred to being "fit" in the sense of being fit enough to stay in a sustained relationship with the available resources. "I use this term [Struggle for Existence]," Darwin said, "in a large and metaphorical sense, including dependence of one being on another." The idea that fitness means being strong enough to oppress or destroy the weak is a fantasy, fueled by fearful humans who are certain they live in a dog-eat-dog world demanding that they fight their way forward.

Yes, it's true, as even the ancient Taoists understood, that the natural living of one thing—its *ziran,* which can be translated as "unfolding"—may curb the lives of beings lower on the food chain. An elk herd, for example, may limit the unfolding of grasses and flowering plants. But if they totally wipe out that food source, their own lives will come to a screeching halt. Likewise, any wolf pack inclined to take down every elk would in short order starve to death. And likely, too, humans who build their success solely on the oppression of others may one day find their own feathered nests collapsing. We live in a world of natural consequences. And those consequences protect the essential vitality of the whole.

Out in the nonhuman natural world, especially aggressive individuals may end up not on top, but yanked from the gene pool altogether. Two aggressive red deer bucks may be so into fighting with each other during the rut that they're too distracted to see that other males are busy breeding the very adult females they're fighting over. Overly aggressive alpha baboons are sometimes less successful when it comes to breeding than are lower-status males who cultivate strong friendships with females. An increasing number of scientists, from biologists to geneticists to anthropologists, have suggested that we may want to start defining overall health—be it in nature, the human body, or society—not in terms of competitiveness, but by the degree of established cooperation.

In my decades roaming the Northern Rockies, I've often watched how the mighty grizzlies spend less time taking down prey animals than they do eating moths and ants and tubers and chokecherries. To some people, that can be downright disappointing—jarring even—a shattering of an invented world where big, strong bears are supposed to be creatures against which we celebrate our own toughness. Writing for *Harper's* in 1872 about grizzlies in the high country of Colorado, hunter Colvin Verplanck was hugely disappointed to discover the great bear dining on lowly grasshoppers—a sight, he later wrote, that caused the creature to "dwindle in our estimation." Offering a reality check, one of Verplanck's contemporaries, the great naturalist John Burroughs—from 1870 to 1920 the most popular nature writer in America—said something rather different about survival. The challenge of life, he noted, is only "the struggle of the chick to get out of the shell, or of the flower to burst its bud, or of the root to penetrate the soil. It is not the struggle of battle and hate."

Then there's that hugely popular adage, especially common in the United States, which is to see someone as a "rugged individual." But there's a problem with that, which is that you can wander the Earth, plumb the oceans, traverse the ice packs, and nowhere will you find, not ever, such a thing as a self-contained being. An alpha male of a wolf pack picking on a healthy bull elk instead of joining his group to go after more vulnerable prey would end up dead before his time, his skull smashed or his ribs kicked in. Not good for him. Not good for his family. Not good for his species.

We creatures of this planet are, every single one of us, fully and inextricably interdependent. The lion didn't learn to run fast in a vacuum. She mastered her speed because the impala she hunts for food learned to run a little faster first.

We've held on to notions contrary to the wisdom of nature not because we're stupid. But as science became ever more popular, a kind of oracle for ultimate truths, we often wrapped our lives in popular scientific views. Sometimes we just misinterpret the science—like with "survival of the fittest." And sometimes that misinterpretation is calculated, used for nefarious purposes, as when so-called social Darwinists of the late 1800s applied survival of the fittest to factory workers sickened from terrible working conditions—dismissing them as weak, their premature deaths simply "the way of nature."

There are lots of far less insidious examples of people overpainting their lives with the popular science and technology of the day. By the time Descartes came along, the world was a good hundred years into being driven by gears and levers and springs. Which led people to think of the universe—right down to the brains and bodies of mammals—as machinery. Likewise, when chemistry was on the rise in the eighteenth century, many came to see human life and love in terms of chemical reactions. Then, when the first breakthroughs in wireless communications happened a hundred years later, the brain was suddenly being compared to a telegraph. In the 1950s we became computers.

And when science insisted in the seventeenth century that hyperrational thinking was the only true form of intelligence, that idea, too, came to roost in every corner of society. Back then in England or France, it would be hard to miss the haughty dismissal among educated people of any art or literature that used symbolism. It wasn't just unpopular, but downright offensive to make art that didn't bow to the purely rational mind. What had come before in the way of poetry and story, including those fanciful folktales of fairies and pixies and nymphs and wildland spirits, was dismissed out of hand. As for the tales of Africa, or even the rich, layered stories coming out of the Americas, like those of the Penobscot and the Iroquois, they were quickly dismissed as the sorry products of pathetic, inferior minds.

In the worst cases, the misappropriation of science can spin utterly out of control. As men like Descartes and Pierre Gassendi and Thomas Hobbes set off on their mission to disenchant nature, European religious leaders were quick to join. With religion increasingly seeing the natural world as the hiding place of the devil, disenchantment became a holy endeavor. That frenzy swelled until it initiated what came to be known as "the Burning Times," when some fifty thousand women and ten thousand men were declared to be witches and pitched into the fire.

Thankfully, throughout such times there have always been outliers among the powerful. As there are today. You may be one of them. People who resist the excesses roiling through the streets from whatever the current big thing of the day happens to be. In Enlightenment-era Europe one outlier was Italy—a country that while greatly influenced by logic refused to lay down its art on the altar of rationality. As a result, other countries declared Italian writing hopelessly inferior, even as it was giving birth to stunning creations from poets like Giuseppe Parini and Gasparo Gozzi. Dismissed, too, were the thoughts of Italian philosopher Giambattista Vico, who himself was no stranger to the precision mathematics that were driving scientific thought. Vico warned repeatedly against training young people in analytical behavior alone, saying that such overfocus would blunt their imaginations. Few in the rest of Europe bothered to listen.

Italy's refusal to go headlong into hyperrationality may have also left it more open to the creative contributions of women. The brilliant Italian scholar Laura Maria Caterina Bassi earned her degree of philosophy in 1732 and in that same year defended a whopping twelve new theses. Though she wasn't given the same free rein as her male colleagues, by 1760 she was nonetheless earning one of the highest salaries of any academic in the country, going on to become a much-celebrated professor of experimental physics at both Collegio Montalto and the Institute of Sciences. All of this while fabulously learned white men to the north, in England, were busy concluding that superior male intellect was a fact of nature.

Those who balked at this epic drive to disenchant, who failed to champion the notion that there are eternal truths discernable, as Voltaire put it, by "anyone of good sense," were considered ignorant, undeserving of respect or consideration. So should you ever wonder where hostility toward intellectuals comes from, or why educated white men have so often irritated the crap out of people from other backgrounds, the Enlightenment is a good place to start looking.

Science itself is now a good hundred years into breaking free of its old insistence on pure objectivity. The end came with a hammer, and that hammer was quantum physics. Quantum physics shredded the deeply held idea that nature could be dissected and held fast by objective observation. Besides that, it brought to light the fact that traditional physics and the mathematics it rested in weren't an infallible means of prediction after all. And that was nothing short of earthshaking.

Early on in the quantum era, through something called the "double-slit experiment," it was found that when it comes to measuring the behavior of electrons, it's the act of measuring that determines their location. A person looking at a thing affects what aspects of that thing are actually seen.

Quantum physics taught us that if we set out to prove light energy consists of particles, that's what we'll find. On the other hand if we set out to prove light consists of waves, well, we'd find that to be true instead. Light is both—or more accurately, the potential for both—at the same time. This both/and condition was then followed by the discovery of so-called quantum tunneling, which is when particles hit impenetrable barriers and mysteriously pass through them to appear on the other side. And there you have it. On the subatomic level—the very foundation of all life—there's little in the way of ultimate reality, or at least things are far less certain than we long imagined. What exist instead are myriad potentials, all present at the same time. The world isn't fixed at all. In fact it's a rather fuzzy and mysterious place—driven less by ultimate truth than by phantom relationships and shifting potentials.

The cork was out of the bottle. And it wasn't long before scientists started wondering whether quantum behavior could be found in biology too. Even today quantum explorations are knocking the legs out from under a lot of biologists who believed they knew the nature of nature. Chlorophyll alone stirred things up. We now know that the moment a chlorophyll molecule captures a photon of sunlight, it sends that photon to a kind of processing center where it gets turned into chemical energy. But en route to its conversion, the photon doesn't just follow one path. It explores many pathways *at the same time,* adjusting instantly—without any discernable passage of time—to locate the path of least resistance. Sure, you can ask the question: "Where, right in this instant, will I find the photon?" But if you expect the answer to follow the rules of classical physics, which tells you it can be pinpointed—well, that's when things start to fall apart.

We're now facing the impossibility of determining unequivocally even what of the world belongs to you, and you alone, and what doesn't. Or for that matter what belongs to a turtle, or a zebra, or a bacteria cell. Animals, plants, bugs, and humans are not independent, not at all, which means they can't really be defined by precise boundaries. Rather, everything, much as Thich Nhat Hanh said of that sheet of blank paper, is a complex weave of relationships and porous, overlapping borders. Everything, in other words, contains everything.

Thanks in part to quantum science, were you to go to a lecture today at Utrecht University in the Netherlands, maybe to hear the brilliant physicist Jim Al-Khalili, it would be a very different experience than you would've had listening to Descartes four hundred years ago. You'd likely find Al-Khalili talking about the world in astonished tones, thrilled with the mystery of it, calling the processes of quantum biology "almost magic."

Instead of a machine doctrine, we'd hear a breathtaking vision of shared reality. We'd have a remarkable sense that the more we know, the more we don't.

"Science cannot solve the ultimate mystery of nature," conceded the brilliant Nobel quantum physicist Max Planck in 1932. "And that is because, in the last analysis, we ourselves are a part of the mystery that we are trying to solve." Just his saying that was really big. So big that ninety years later a lot of us are still trying to get our heads around it.

We're many years out from pure objectivity being held up as the one true way. Yet when I go outside tonight to look at the stars and planets, my initial sense will be that all those beautiful shimmering points of light—Orion and the Big Dipper and Mars and Jupiter—are incredibly lovely things "out there," they in their world and I down here in mine. It will take a little while, a little settling down, before I can really *feel* the fact that the force of creation that gave those stars and planets their mass and their orbit and their chemistry is the very same force, with the same materials, that produced an Earth with such a staggering flush of life. When I'm successful, when I can manage to go beyond seeing disparate pieces and rest instead in what's really a vast sea of connection, then—and only then—does the often-quoted maxim that we're all made of stardust go from poetry to reality.

In the early 1990s I was in upstate New York, doing research for a delightful book project about people whose lives were still firmly tied to the beauty and lessons of the American forest. On a day off—a welcomed time, spent rambling aimlessly in my old van—I found myself untangling the highways along the Hudson River. Not really looking for anything; just making right turns here and left turns there, looking for something to grab my attention. The sun was pouring onto the crumpled loaves of the Catskill Mountains and lighting the Hudson in a breathtaking flash of silver.

But it was close to rush hour and lots of other people were on the road, none of them in the mood to be behind a gawker from Montana driving like he'd just fumbled his glasses out the window. The honking horns, along with a few other, stronger gestures, prodded me to find a turnoff. The first opportunity was at a narrow wooded road leading to a place called "Olana." Which is how I found myself on a high bluff above the Hudson River, at the magnificent Persian-style mansion of the nineteenth-century landscape painter Frederic Church—for a time the most famous artist in America.

The museum was closed, but museum director Jim Ryan was kind enough to invite me in to see a few of the rooms. When he talked of Frederic Church and of the famous Hudson River school of landscape art in general, his voice, the sweep of his arms, had the easy comfort of old books and stuffed chairs. I began imagining myself listening in a smoking jacket with a snifter of brandy.

The attention to detail in the mansion was captivating. Every room was filled with art collected from around the world, from China to Assyria to Mexico. Interior walls carried paints and designs in which every color was painstakingly mixed by Church on his own palettes. All the while, wide, green views were pouring in through every window. In the sitting room was a final sketch for Church's *Twilight in the Wilderness,* considered by some to be among the greatest works of art in American history. It was a scene thick with nature, painted on the eve of the Civil War. And because of that, the whole of it was dark and red, holding its breath, as if every good thing was about to drift from our grasp.

Strolling past this and other paintings in the murky light of the closed museum, including some created by Church's brilliant contemporary Thomas Cole, sent me spinning back to my boyhood. These were the exact images that at age eleven had set me dreaming of bigger nature, the whole idea of wildness jumping into me from the pages of the art books I fawned over every week in the South Bend Public Library. These Hudson River school artists passed on to me the myths and longings of a whole culture; I inhaled them deeply, and in little time made them my own.

But there was something else in their art that in the end steered my experience of the world around me, a major regret: namely, that we Americans had through our neglect and abuse of nature placed ourselves outside its sacred circle. We'd come into a genuine Garden of Eden all those years ago, said the Hudson River painters, and through our own greed had kicked ourselves out. The ravaged, overcut forests of New England these painters saw in the mid-nineteenth century, the resulting ruined watersheds and fisheries, were to them an affront to God. Many, like Thomas Moran and Albert Bierstadt, headed west to focus their attention on less spoiled landscapes. And as the images of nature on their canvases got bigger, the human figures those canvases contained, at least the Anglo figures, shrunk until they were barely perceptible.

It proved an enormously powerful notion, this sense of being forever estranged from the Earth, of being pushed out of the garden. And it remains a cornerstone of how we see ourselves today. Good, beautiful, wild places over there, greedy humans over here. In my own years as a young man, riled up by everything from Three Mile Island to Love Canal to smoldering toxic dumps in East Chicago, I embraced much the same idea. Church's home at Olana, high above the Hudson, though astonishingly beautiful, felt on that summer day like a monument to my own anger and sadness.

However, the idea of kicking ourselves out of the garden as a kind of self-punishment is itself a trap, a kind of binary thinking, which, in its own way, is an act of separation no less objectifying than the Greek idea that nature could only be studied by standing outside of it. We don't get to throw ourselves out of the garden. We remain connected because that's the only way any of us gets to live on this planet. We can grieve our mistakes, take a couple of deep breaths, and get busy patiently repairing the relationships we've either strained or left in tatters.

Given what's coming to light about the critical importance of interdependence in nature, it probably shouldn't be too surprising that some of the oldest stories on the planet have been about that very thing. This one, from Java, Indonesia, is just one of hundreds, each with its own cast of characters based on its location in the world. But what doesn't vary is that the characters are always caught in the same dilemma.

Forest and Tiger were the best of friends. They talked of this and that, but mainly they just liked being together, watching the coming and going of the seasons, tracking the roll of sun and moon as they slipped across the sky, high above the branches of the trees. With the coming of humans the two friends came to appreciate each other even more. You see, people were reluctant to come and cut down the trees because they feared being eaten by Tiger; likewise, when humans came with spears determined to hunt down Tiger, it was the shadows of the forest that allowed him hiding places in which he stayed free from harm.

But as sometimes happens even with good friends, the day came when these two began taking each other for granted. In time this led to feelings of resentment, scorn. "You do little here with your presence but foul my beautiful floor," Forest said to Tiger one day. Tiger responded in kind, scoffing at the forest for its dark and gloomy thickets, saying how troublesome such a place was, how difficult it was to warm his back in the fingers of the sun. Things kept getting worse, and one day Tiger simply walked away, out into the world of open hills and valleys.

Sadly, it wasn't long before humans discovered that the forest was no longer guarded. They rushed in and cut down trees like there was no tomorrow, until the once beautiful forest was nothing but a barren, lifeless place. And Tiger fared no better. Because he was unable to hide in those shadowy thickets, the men with spears found him easily, killing him and all his kin. And thus what had at first seemed like nothing more than an unfortunate parting of ways between friends was in fact the beginning of the end—the undoing of both Forest and Tiger.

The early storytellers who told this tale were well aware of the importance of relationship in nature, of the need for balance. And despite the fact that we've come through centuries of men claiming the contrary, objects are not now, nor have they ever been, more real than relationships. That tree, that cow, and that person across the street with the red boots and black umbrella are unique expressions of the flowing patterns that buoy us all. To engage any of it is to engage the world.

Embracing that fact means making room for a kind of "beginner's mind," to borrow a concept from Eastern meditation practices. The rest of the shift comes from putting ourselves in situations where the sensory input we're getting is bigger than we can comprehend. A matter of putting down the need to catch reality in a glass.

And that's where nature comes in. Deep experience of nature can cause powerful perceptual changes in part because it allows what our brains anticipate about daily life, about how things work and what will happen next, to be gently overwhelmed. In a sense we're given more color and light and movement and sound and smell and connections than we can possibly process. Instead of being a source of anxiety, though, this kind of feast can actually lead to a softening. It can guide you into a place where not only do you not know what's going on around you, but you're actually pretty pleased not to. The effect is to create a fresh canvas onto which entirely new, more relational ways of thinking can be painted.

As we begin moving more deeply into acknowledging the role of interdependence for life on planet Earth, it's important to note that this also means being connected to other people. Much of who we are, after all, has been shaped by interactions with those we love, and at the same time, by whatever walls we may have erected against those people or groups we don't like, or simply don't know. It's especially fascinating to consider that research in the fledgling field of epigenetics suggests that our own life experiences may rest to a degree in the experiences of our ancestors. Not just physical traits, mind you, but things like profound trauma appear to be passed across time through our genome, reappearing as the potential for certain behaviors or even illnesses in every new generation. Holocaust survivors, for example, tend to have lower levels of the hormone cortisol—a trait that by all indications seems to be passed along to their offspring, creating a potential risk factor when it comes to high levels of stress.

What if, however, as some psychologists are starting to wonder, it might also be true that to the extent we experience profound happiness in our lives—to the degree we use those positive conditions to heal our fear and our pain—this resiliency, too, might express itself in future generations?

As we begin this journey to better embrace the facts of connection, we might do well to think about the old notion of *ubuntu,* from the Nguni Bantu people of southern Africa. *Ubuntu* acknowledges that all of humanity is linked first and foremost through a bond of sharing. And further, that true wealth, true satisfaction, can never be fully realized if our own prosperity isn't shared with the others with whom we travel the world. On one hand such a notion echoes the primary doctrines of virtually every religion on Earth. And at the same time, on a cut-and-dried practical level, it's about living in a way that ensures the welfare of our grandchildren—through the health of not just the external world we pass along to them, but the internal one too. Healing into our lives, as countless poets and artists have tried to tell us over the ages, means learning to see life as a creative, passionate, messy, lighthearted dance. And not a solo dance, but a dizzying, dynamic waltz with a spectacular array of both human and nonhuman partners.

Returning to poet Walt Whitman: Looking to heal from the stroke that had paralyzed half his body, he was at the same time still reeling from the work he'd done as a nurse on the nightmare battlefields of the Civil War. Whitman took to living for a time in a patch of New Jersey forest known as Stafford Farm on Timber Creek. Soon after arriving he began waxing in his notebooks about a certain old tree he found terrifically comforting.

"How strong, vital, enduring! . . . What suggestions of imperturbability and *being,* as against the human trait of mere *seeming.*"

Science, he went on to say, tended to scoff at the way some people reminisce about spirit nymphs of the forest, or even the idea of trees speaking to us. But those reminiscences, he insisted, were quite as true as any other perspective, and more profound than most. Go and sit in a grove or woods, he advised. Sit with one or more of those voiceless companions, and just think.

"After you have exhausted what there is in business, politics, conviviality, love, and so on—have found that none of these finally satisfy, or permanently wear—what remains? Nature remains."

As Whitman suggests, it's unwise to ignore the value of our waking dreams, disparaging our flights of fancy as trite and imprecise, not worthy of serious attention. To do so tears the wings off a relationship with the Earth thousands of years in the making.

Curiously, attaining this broader, more imaginative kind of intelligence appears to also grow our capacity for sympathy and compassion. And that's no small thing when it comes to mental health. Darwin called sympathy not only one of our most valuable instincts, but our strongest. Recent research by Elizabeth Nisbet and John Zelenski shows that the level of emotional connection people have with nature actually predicts the likelihood of their establishing healthy, contented, productive lives—one aspect of which turns out to be tolerance and love of others. When a person dismantles the wall between herself and the natural world, what often rushes in is the pleasure of reunion.

Some of us in the dominant culture have been prone to assuming that indigenous cultures living in places for thousands of years kept going in large part because their technologies were so limited that overharvesting resources was impossible. To be honest, I grew up thinking that very thing. If they'd had the extractive machinery we have, the presumption goes, they too would have taken all they could.

But almost certainly, the longevity of those cultures has a lot more to do with practices of balance grown from daily exposure to the natural world, living every day deeply embedded in the experience of being linked to "the other." Living in this way likely left them far less willing to disrupt the web, to tarnish the blessing of life at large. They had only to look at the world itself to nourish a conviction that by disrupting the balance they would risk terrible loss, ripping from their lives sustenance and comfort and quite possibly unwinding society as a whole.

When writer Sylvia Plath was just nineteen years old she wrote in her diary of an afternoon spent at a grand stony beach, where she climbed to a large rock jutting out over the sea. From that vantage point she contemplated the tides rising and falling, the silent drift of sailboats out on the far horizon. She considered how water and weather had sculpted and scoured and broken the rocks along the shore. There was the sound and feel of wind fingering the nearby grasses, tossing her hair. Like so many who go to beautiful places in the natural world, she came back transformed, with fresh eyes with which to see her own life.

From such experience also, she explained, "a faith arises to carry back to a human world of small lusts and deceitful pettiness. A faith, naïve and childlike perhaps, born as it is from the infinite simplicity of nature. It is a feeling that no matter what the ideas or conduct of others, there is a unique rightness and beauty to life which can be shared in openness, in wind and sunlight."

I know something of this faith. From watching ten

thousand pink flamingos settle into Ngorongoro Crater in Tanzania to catching my breath at the sight of caribou dancing like trotter ponies across the broken tundra of the Far North.

But one of the most powerful days came halfway through a five-hundred-mile walk around the perimeter of Yellowstone. I was in the lazy swales of the magnificent Bechler River valley, in the southwest corner of the park. Having just set up camp, I grabbed my tape recorder and journal and wandered down to a grassy seat at the edge of a copse of trees on the west bank of the river. There, in a lacy patch of yarrow, I waited. In the arid West, after all, sooner or later nearly every living thing makes its way to the river. It was late summer, the sky a flawless blue. The air was filled with the smell of pine and cured wheatgrass. Beneath me, along five gentle bends of the river, countless rings spread across the surface of the water as trout rose to feed on mosquitoes— and just above the surface, tree swallows, zigging and zagging to nab insects on the wing. In the meadows to the west I could hear the fluty, primeval chortle of sandhill cranes.

Before long a fine young cow moose ambled down to the far shore, crossing over to begin a casual feast of willow and pondweed—staying there for a good hour before finally swaggering off into a loose thatch of forest. As it turned out she was the grand marshal of sorts, kicking off what seemed like the Mardi Gras of animal parades. First a female merganser drifted past with her brood of young chicks. Then, a hundred yards behind, a family of mallards, led by one of the most indomitable mothers in all the bird world, drifting with the current through soft green ribbons of aquatic grass. And then an osprey on the wing, cruising the river looking for fish, the air making sounds like whispers as it rushed beneath her wings.

I was getting hungry by then, but the river held me fast. Next came a family of beaver swimming upstream against the far bank, pausing just thirty yards or so beyond me to frolic a bit—one of the adults slapping her tail with a loud crack, water splashing all around, not as an alert to danger, as it can often be, but more part of what seemed to be the beginning of a game of chase, the whole kit living as if danger were days away. Barely had they moved on when an otter appeared—one of my favorite creatures—slipping under the water and coming up with a cutthroat trout, then rolling over to eat it off his stomach, glancing at me every now and then as if to show me a chapter of the good life according to otters.

By this time it felt rather like I'd fallen through some kind of cosmic crack. My tape recorder was on and I was whispering the details of every creature's passing, and even today I still smile over the thrill in my voice when I listen to the tape of me describing the unfolding of these events. Ten minutes later came more visitors. Swimming at the far edge of the river, nibbling on sedges and the bulbs of various aquatic plants, a family of muskrats.

I knew then that even if I never set foot in that landscape again I'd rest easier knowing it was there, much the way we can take solace at the thought of composers out there creating new music whether or not that music ever falls on our ears. I was a more centered person in those sweet two hours, gentled and heartened by a display of life so rich and complex that the only thought worth having was a kind of silly senselessness at the wonder of it all.

Twenty-five years later I'm more convinced than ever that what nature has the power to do—even in far smaller doses than what I saw along the Bechler River—is to release us for a time from the tiresome obsession with self. In those first few moments by the river I was a hiker calculating the miles I'd traveled. Savoring my physical prowess, shoring up my own story. But the swaying hips of the Bechler, the tender colors of the meadows, and the graceful swimming of the beaver and the muskrat coaxed me up and out of myself, rooting in me a sense of the tremendous, unfailing *reliability* of nature's relationships. The sort of liberated reality that somehow makes it safe to move, if only for a little while, out to the margins of our lives, to open our arms and call for something bigger than what we often see in more common hours.

# The More Kinds of Life in the Forest, the Stronger That Life Becomes

*What is desired is the kind of diversity that fully respects the values and interests of all.*

—Sandra Harding

When I was twenty-one, I worked as a naturalist for the Forest Service in the Sawtooth Mountains of Idaho; one day in late June I was out with my boss and mentor, Chuck Ebersole, a sixty-two-year-old retired navy veteran. Chuck was brilliant, a little ornery, and when it came to nature effusive beyond belief. When I first arrived in the Sawtooths he marshaled me onto a string of orientation tours of the Sawtooth Valley, the two of us rumbling about in his government-green Chevy Vega. On several occasions while speeding down the highway he'd look out the side window, slam on the brakes, and leap out of the car to run off into a nearby ditch or meadow or patch of forest, urging me to follow, talking in exclamation points, wanting to show me something amazing about a flower, a tree, or a slab of ancient stone. It was the sort of frantic behavior most people reserve for spotting bags of money. But to him, everything in nature was a teachable moment. And by the end of his lessons his hands and teeth were clenched, his eyes blazing with barely contained euphoria.

One day Chuck announced that we needed to hike out to one of his favorite mountain meadows. As usual I didn't have a clue what he was on about, but I grabbed my day pack, tightened the laces of my boots, and filled up my water bottle, and off we went. After some two hours of walking we crested a final alpine rise and my jaw dropped. Before me was a tumbling mountainside bedecked with the most glorious carpets of wildflowers I'd ever seen: penstemon and geraniums and buttercups, cinquefoil and paintbrush and bluebells, elephant head and prairie smoke and monkey flowers. For a long time we simply stood shoulder to shoulder in silence, me feeling like my life was unfolding in the middle of a Monet painting. Finally Chuck began an exchange worthy of Socrates, patiently showering me with questions.

*So, Gary. Why do you suppose there are so many kinds of flowers here?*

*Well*, I ventured, *because the soil conditions, the moisture, the light, are just right for lots of species?*

*Yeah, okay. But why all the species? Why not just two or three?*

That one had me stumped.

74

*Think for a minute. What sorts of things might threaten this place?*

*Disease,* I ventured. *Insects. Drought.*

*Okay, let's say it's a big drought. What would happen to these meadows?*

*Well, some of the plants would die.*

*Why not all of them?*

*Because some have deeper roots. Those might be able to survive. Some have waxy leaves to keep the moisture in.*

*Good. And what's the big deal if they do survive?*

*Um, this is a steep slope, so the more root systems that make it, the less likely the soil would erode.*

He nodded but wasn't satisfied. *What else?*

Right then a big horsefly landed on my cheek, and I swatted it away.

*That's right. Flies,* he said with a grin, though I hadn't spoken a word. *If some of the vegetation survives, the flies carry on. And that spangled butterfly over there.*

*And the elk too?* I offered, pointing to a pile of droppings at our feet, venturing a guess that elk, too, might still be able to graze even if some of the plants were gone.

*So what are you telling me?*

I stammered. *Well, I guess I'm saying that different flowers have different survival strategies. And that means that if some get wiped out others will survive.*

*And?*

*And because they survive, the system survives.*

*So you're saying nature hedges its bets.*

I nodded. That sounded about right.

He took a breath. I took a breath.

Standing hand in hand with the fact that we're in relationship to pretty much every other living thing on Earth is this giant and indisputable lesson: The more players there are in a natural system, the more vibrant those players will be. And also, the more resilient the system will be in the face of change. And in this world things are always changing. This beautifully rich and robust planet is, in all seasons, nothing if not a constantly unfolding testament to the essential power of diversity.

The more than one trillion species of plants, animals, insects, and microbes on Earth make up an astonishingly dynamic, deeply connected living layer. A layer that by means of a complex exchange of chemical processes joins together and sustains the atmosphere, the landmasses, and the oceans, seas, rivers, and streams. It's no exaggeration to say that diversity is responsible for giving us breathable air and drinkable water. Not to mention replenishing the soils in which we grow crops, along with creating the pollinators needed to fertilize them.

Thanks to photosynthesis in everything from the trees of the forest to the grasses of the prairies to the phytoplankton in the ocean, we've got the oxygen we need to breathe. Meanwhile plants of every stripe sequester carbon—critical for a healthy atmosphere. Still another big range of plants serve as pollution filters, including willow and alder and iris and cattails. And deep in the dirt, out of sight, is an enormously varied collection of microorganisms that make nutrients available to keep these plants alive and thus allow the world to sustain itself.

Here, then, is an embarrassment of riches. So far we've discovered more than eighty thousand edible plants, and that number is still growing. Diversity has given us everything from the fibers in our clothes—cotton and hemp, silk and wool and flax—to the petroleum that fuels our cars, much of which was derived from a diverse array of ancient marine life including zooplankton and algae, buried and pressurized under layers of sediment and rock.

And then there are the thousands of lifesaving medicines we've come to rely on, with nature even today directly providing some 40 percent of the world's modern pharmacy: Coumadin to help people avoid heart attacks and strokes, originally made from fermented sweet clover; aspirin, first extracted from white willow; the primary drug used to heal childhood leukemia and once-fatal Hodgkin's lymphoma from rosy periwinkle; antibiotics and cholesterol-lowering medicines from microbes; common blood pressure drugs from the Brazilian pit viper; AZT for treating HIV-AIDS from a compound in marine sponges; diabetes medicine from the Gila monster. Spider silk for creating artificial tendons and ligaments, cone snails for epilepsy, coral for cancer treatments. Even the blood of horseshoe crabs (extracted without killing the crabs), which to this day is a critical tool hospitals use to stem deadly infections. Maybe more fascinating still is that recent research shows human rates of respiratory infection, malaria, and even Lyme disease are lower in people living near generally intact protected natural areas. So beyond extracting specific medicines, it's quite possible that the mere presence of vibrant, diverse nature can be a key to our physical health.

When we consider diversity as the main predictor of whether a natural system will be resilient, we see it's not just a matter of surviving over time, adapting to the changing circumstances that are hallmarks of a dynamic planet. It also has to do with how well the community can recover after sudden upheavals like fire or flood, disease or insect infestation, hurricane or drought. It's what Chuck Ebersole taught me high on that mountain in the Sawtooths all those years ago: that if calamity strikes a particular groups of plants, given enough diversity the disruption is highly unlikely to take the whole system down. And with the system still in place, those species that did suffer losses have a far better chance of coming back again.

Much the same is true for animals, where even diversity of behavior or personality can be a boon. A hive of bees, for example, will benefit from having individuals that focus on seeking nectar rapidly, from an assortment of flowers, some with nectar and some without—and at the same time having another set of workers that are slower but much more discriminating. The world of dippers, meanwhile, is made more stable by the fact that some birds live in their home stream as yearlong residents, while their neighbors choose to migrate to winter grounds—preferences, by the way, not necessarily linked to the behaviors of their parents. And while an aggressive bluebird may be just the ticket when it comes to securing territory, his more sociable cousins fare better when it comes to reproduction.

The essential nature of diversity and our utter dependence on it explain the considerable anguish in scientific circles about the so-called Sixth Extinction, which refers to a massive die-off of species due in large part to human-caused climate change. The National Academy of Sciences recently described the Sixth Extinction as "a frightening assault on the foundations of human civilization." And lest you think we humans will be okay because, well, we're really smart, keep in mind that we're part of a primate group prone to extinction. In just thirty thousand years we've gone from several species of humans—beings more or less like us—to just one.

So how did all this diversity happen in the first place? Why would there naturally come to be so many ways of doing the same thing—collecting sunlight and water and nitrogen and turning it into vegetation? Why wouldn't nature's penchant for efficiency show up with a smaller range of options based on two or three really successful strategies?

Many evolutionary biologists now believe, as Darwin predicted, that the three domains of life on Earth—bacteria, microbes known as archaea, and multicelled creatures ranging from plants to animals to humans—evolved from a single-celled organism that lived about 3.5 billion years ago. Like those original life-forms, every life since has been composed of the same twenty-three "universal proteins." That shared characteristic has anchored the evolution of living things across billions of years. In that way, all of us are indeed deeply related. And at the same time, this foundation of twenty-three proteins, together with the dynamism of countless evolutionary strands, has yielded the wide range of life we see today.

Still, what explains such a big range? How did so much diversity come to be? One small doorway into answering this question can be found in the tango that's been going on between plants and insects for more than four hundred million years. And I do mean tango, each one greatly affecting the other through a push and pull that's driven everything from slight changes in the organisms to the appearance of new species.

One strategy many plants have honed over great leaps of time for holding plant-eating insects at bay involves producing ingenious defense chemicals—called secondary compounds—in their leaves and flowers. For example, dozens of plants, including tea and coffee, make caffeine for that very reason. Oil in the leaves of mustard works the same way. Meanwhile, milkweed deters would-be leaf munchers by producing poisonous glucosides. This dodge to avoid becoming a choice entrée for insect dinners helps milkweed do better in the world. By protecting its leaf tissues, the plant grows bigger. It also produces more seeds. And the more seeds the milkweed makes, the better it will be at propagating itself.

But producing defense chemicals isn't the end of it. Over time certain insects or caterpillars, like those of the monarch butterfly, can manage to overcome the problem of a poisonous plant often through a mutation that yields a sort of superpower, allowing them to no longer be affected by the toxin. This is how the monarch butterfly can go about its business flaunting its bright colors for all the world to see. Having gained that ability to eat milkweed, the monarch turned itself into a noxious meal. Any insect-eating bird that might be sitting in a nearby tree watching the butterfly float about within easy reach will pass it by because the bird knows eating monarchs will make it sick. Moths and butterflies that haven't gained the monarch's superpower will often opt instead for camouflage, fashioning wings that look so much like the vegetation they live on that birds can't spot them.

All in complete concert. No new species ever rises by itself. The plant pushes the insect in new directions, sometimes helping foster new species, and the insect pushes the plant. The tiger pushes the antelope, and the antelope, gaining new behaviors and abilities over time, pushes the tiger. In that sense we're back to Lesson Two—interdependence. Diversity rises, in significant part, from relationships.

So can we draw a line from this lesson—that diversity is the key to both biological creativity and sustainability—to our own human communities? As it happens, countless people, many outside the scientific community, have thought long and hard about that question. John F. Kennedy expressed hope that while people might never be able to end their differences, prosperity depended on making the world safe for diversity. Maya Angelou advised parents to teach their children early on "that in diversity there is beauty and there is strength." Even businesspeople have seen the power of diversity, among them Malcolm Forbes, who defined it in business as the art of thinking independently together. Appropriately, the Latin word for thinking, *cogito,* comes from a root term that means "to shake together."

One of the more intriguing thinkers on diversity in human communities was a feisty mid-twentieth-century writer, economist, and urban activist by the name of Jane Jacobs. Born Jane Butzner in Scranton, Pennsylvania, as a young woman she began working as a freelance writer. Much of her focus was directed toward life in the city—in particular, her chosen home of Greenwich Village. She was dogged and patriotic—a big believer in independent, uncorrupted democratic city governments, tax-supported public services like mass transit, and always, smart rules and regulations to help preserve diversity. Like her contemporary Rachel Carson, Jacobs was routinely ridiculed because she was a woman—in her case by the male bastions of city development and planning. But no matter. When it came to spotlighting the impacts the wheels of progress had on poor and disenfranchised people, she was fearless. And to shore up her arguments, she turned again and again to the natural world.

When Jane Jacobs launched some of her most important work in the 1950s, science, industry, and government were overflowing with a muscular, bully brand of hubris. These were the days of nuclear fallout—so-called black rain—dropping from the skies and spiking breast cancer and thyroid disease in thousands of Americans in the West and Midwest. Meanwhile we were well under way to spraying more than a billion pounds of DDT on farmlands and city parks—something Rachel Carson would blow the whistle on in a big way in 1962 with the publication of *Silent Spring*. The 1950s saw dramatic rises in levels of mercury and lead in the nation's air and water, as well as enormous increases in smog. Indeed in New York City alone, three separate smog events between 1953 and 1966 killed more than six hundred people.

The size and ferociousness of the industrial juggernaut were breathtaking. Far to the west, the Bureau of Reclamation was pushing hard to shove dams across many of the last wild rivers, including ill-fated attempts to plug Colorado's Green River in the heart of Dinosaur National Monument, as well as the Colorado River in the Grand Canyon. Public forests throughout the Pacific Northwest and California were falling to the saw under massive clear-cutting, including the destruction of nearly all the giant sequoias still on private land. Wetlands were being filled by the thousands. The government was still sponsoring bounties on a wide variety of "bad" animals, from cougars to coyotes, wolves to weasels, hawks to owls. Progress was measured in explosions and chainsaws and poison and wrecking balls, and no one who dared stand in the way had an easy time of it.

While writing articles about cities in the Northeast, Jacobs saw firsthand the devastating impacts of steamrolling entire neighborhoods to make way for the massive multi-lane freeway systems central to so-called urban renewal. But again, what makes her fascinating is how she came to forge powerful arguments to halt those projects for the sake of the city's beleaguered residents by turning to nature—in particular, using what the natural world has to say about diversity for keeping life healthy and resilient. She was among the first to point to the value of diversity in those neighborhoods—the continually erupting street-corner encounters and sidewalk exchanges between neighbors and small businesses, which she affectionately called "the sidewalk ballet." Jacobs revealed the ways the weave of neighborliness kept the city strong, able to reinvent itself in the midst of ever-changing conditions. Segregating people into housing projects, she said, removing the highly complex sidewalk life of the urban neighborhoods, was an "unnatural act" that doomed many residents to lives of poverty and despair and, at the same time, squelched the local economy.

As often as not she started conversations about city planning with talk of ecology. Maybe she'd point out how all life systems are created by the successful capture of power from the sun, and that the more avenues there are in that system to claim the sun's energy and then pass it along—in other words, the greater the diversity of plants, animals, and microorganisms—the healthier and more resilient the system would be. Life grows by harvesting energy, then using it to sustain existing endeavors as well as passing it on to create new ones.

Over and over she invited people to consider how that particular truth of nature mapped onto human economies. "Nature cast us up with these abilities," she said in an interview in her eighties, as lucid as ever. "And they're just as natural to us as the spider's ability to weave webs and to sting prey. Or a bee's ability to make honey. What we can do with our brains, what we can do with our hands is natural to us. It comes from nature. And we use it the same way that nature develops an ecology. This is not a metaphor. It's the same process."

When discrimination keeps the people of a neighborhood from bringing their creativity to the table, Jacobs argued, then the work those people do is essentially sterilized. Destroying the ecosystem of an entire community stopped the very economic base from which new and even more creative enterprises would arise. If whole categories of men and women doing specific work weren't able to use that work to launch new inventions and ways of doing things, the resiliency of the economy weakened.

We don't have the option of developing some other way, she insisted, because the fact is that some other way doesn't exist. Protect human diversity in our everyday lives, respect every person in the mix, allow them a part in combining and recombining ideas and resources—both on the job and on the street—and you end up with the very best chance for a vibrant long-term economy. The success of nature isn't a zero-sum game, where every gain means a loss someplace else. Mature ecosystems, after all, ones filled with all sorts of life, manage to keep still more life coming.

Jacobs saw that nature's ability to repurpose, to take existing designs—from wings and feathers to feet and fins—and tweak them into something new, something more adaptive to emerging conditions, was also what humans were prone to do. A person working for the sanitation department is struck by the amount of used clothing being thrown away; working with local charities, he manages to slot the throwaways into a fledgling new economy—selling used jeans to China, thread to India, tattered remnants to the Midwest to be made into automobile seat covers. A nineteenth-century cotton-mill worker, Margaret Knight, imagines and then invents a sewing machine to make paper bags with flat bottoms—creating a new design standard still used today. Shaker tool maker Tabitha Babbitt has the bright idea that timber cutters could save effort by using a circle-shaped saw blade, instead of one that has to be pulled lengthwise. She makes a prototype and attaches it to her spinning wheel, creating the first circular saw.

"The kind of mind I have is basically a scientific one," Jacobs told an interviewer late in her life. "I respect observation and experiment. I like to know how things really work." Our creativity, our experience and skills and human capital, are wonderful. Like nature, they don't run out. "The more you use it, the more you have of it."

The power of diversity in the human world affects a lot more than just economic resiliency. Say you're part of a group of people trying to solve a tough problem at work. If you and your co-workers are from more or less the same background—maybe middle- or upper-middle-class white men with similar life experiences and educational backgrounds—then you're likely going into meetings assuming, and rightly so, that your colleagues understand the world pretty much like you do. Which means that each of you is probably going to come up with somewhat similar solutions to whatever problem you're working on. And while this kind of familiarity is comfortable, even reassuring, it can also be a block to creativity.

Mathematical biophysicist and science historian Sandra Harding calls this familiarity "weak objectivity." Strong objectivity, on the other hand, arises from listening to people at the margins—those who aren't part of the dominant group—who, by the way, have had to know the rules of the group in charge in order to survive. Related to this, social scientists have found that sameness in working groups tends to lead each of us to be less diligent when we share an idea, less clear when we speak. This is because we assume everything we say is fully understood by our listeners. We also don't hear as well when we listen. When people who look and experience the world like we do are talking, we often don't pay quite as much attention; we bend the words, we "fill in the blanks" with pieces of what we assume is a common vision.

Agility and creativity are fostered by diversity. Is it more work? Yes. Are outcomes better? Most definitely.

Not long ago a team of researchers perused more than a million scientific papers—scholarly articles subject to rigorous peer review—sorting each one by the level of ethnic and cultural diversity in the teams of people who authored them. What they discovered was that papers created by groups with higher levels of diversity had significantly more impact—meaning they ended up being cited more often by other scientists—than papers from homogenous groups.

And we're not just talking about ethnic and cultural diversity. Strength also comes from differences in how we function mentally. In 2013 a major scientific breakthrough occurred when astronomers figured out how to use "flickering starlight" to gain a better sense of the size and evolutionary stages of individual stars. Team leader Keivan Stassun says the accomplishment came in part because he intentionally assembled research groups made up of not only different races and genders, but also an array of intellectual styles. Some of his team were on the autism spectrum. Those members happened to be especially good at intense focus, "drilling down" through a string of questions for long periods of time to uncover critical pieces of the puzzle.

But they also sometimes struggled in large-meeting settings, which can be full of free-ranging, sometimes off-topic conversations. Seeing this, Stassun made it possible for them to participate in those meetings at a distance, texting back and forth with the group.

Stassun says that discoveries like the flickering starlight breakthrough are just more likely when you have people from diverse backgrounds, because diversity pushes a team out of easy assumptions. Science is always better served, he explains, when you have "a group of very different kinds of people looking at the same data from different perspectives."

As Columbia vice dean Katherine Phillips writes, "Simply adding social diversity to a group makes people *believe* that differences of perspective might exist among them—and that belief alone makes people change their behavior." She goes on to say that when people know they're socially different from one another, expectations change. "They anticipate differences of opinion and perspective. They assume they will need to work harder to come to a consensus." And the benefits are far ranging.

Should you ever be unlucky enough to find yourself on trial for a crime you didn't commit, you'd be smart to hope the jury is made up of both people of color as well as whites. According to social psychologist Samuel Sommers, such juries are significantly better at reviewing cases, typically making fewer errors when recalling information revealed at the trial. In the courtroom, much like the boardroom or research laboratory, more diversity leads to better attention and deeper diligence.

Again, we humans arose from the natural world. We're undeniably part of it. It makes sense, then, that diversity would give us strength. That it would give us big advantages in the world at large.

Out in nature the health and resilience of a natural community depends not just on varied species, but just as importantly on the range of traits and behaviors within each species. One Shasta daisy may be better at putting down roots, while another Shasta daisy is better at producing blooms, while still another is best at conserving water. Or when it comes to animals, an aggressive female elephant may be just the ticket when the group is being threatened, able with her size and ferocity to send challengers and would-be predators packing. But the group will also benefit enormously from its quieter elders—those able to build strong relationships within the herd, anchoring the sort of cohesiveness critical to their long-term survival.

Here in the human world, of course, the differences we see within the species come from a wide range of factors. Beyond personality differences, our lives are shaped by the color of our skin, by our age and gender, by the build and function of our bodies. There are differences in where we live, in our religious beliefs, in our sexuality, in the amount of learning we've done both in school and through experience, and also in how our brains process information and stimuli. Which is one reason why humans as a collective have the potential for an extraordinary range of creativity.

Yet we haven't always been good at tapping the power of diversity, even when promising to do just that. Early scientists of seventeenth-century Europe liked to tout the notion that they were open to anyone and everyone who had something to contribute. The pursuit of scientific truth, they said, was a way to break free once and for all of a culture that for too long had favored only the opinions of political and religious insiders. Yet in reality, women, Jews, people of color, and the poor had no access to the education that would give them the math and Latin and botany they needed to get into the science club in the first place. One of the most prestigious scientific communities in the world, the Parisian Académie Royale des Sciences, founded in 1666, would not admit the first woman—she the wife of an academy member—until 1979.

Not surprisingly, this kind of selection bias created a science with a very limited perspective. But the biases would have even more insidious consequences. They had to do with the fact that if we make it impossible for certain groups to join the brain trust, inevitably some members of the club will jump in to explain away the absence of those people as proof of their inferiority. And that prejudice, which has routinely leaked into the culture at large, is profoundly self-replicating.

In the early decades of the 1800s, one of the nation's most respected physicians and natural scientists, Samuel George Morton, embarked on a quest to settle a long-simmering question among whites as to which of the world's races was the most intelligent. Many of his colleagues applauded him for taking on such a project, sure that his skill and honesty were beyond reproach. As the *New-York Tribune* would later write following his death in 1851, "probably no scientific man in America enjoyed a higher reputation among scholars throughout the world than Dr. Morton."

Morton made the baseline for his studies a collection of human skulls—eventually it grew to about a thousand—collected from various peoples around the world, including Africans, East Indians, Western Europeans, and Middle Easterners, as well as both South and North American Indian tribes. The idea was to rank the skulls according to the size of the cranial capacity, which to Morton and many others of his time was thought to be a foolproof indicator of intelligence. To complete the task he filled each of the skulls with mustard seeds, filtered to ensure that all the grains were of equal size. Later he took to using lead shot, the kind contained in shotgun shells.

The groups whose skulls held the most seeds or lead shot were deemed to be the most intelligent. At last Morton finished his research, and against great anticipation announced the results. Western Europeans came in first—the most intelligent. American Indians were in the middle. Africans were on the bottom. And those results formed the bedrock of arguments about race for the rest of the century, including being routinely cited by southern politicians making the case for slavery.

But beside the fact that cranial capacity wasn't a reliable means of gauging intelligence in the first place, there were other problems. Big ones. In the late 1970s renowned paleontologist Stephen Jay Gould went over Morton's research notes with a fine-toothed comb. He found Morton's methodology sloppy; in fact, the age and sex of many of the skulls wasn't even known. But Morton's data were also riddled with strings of guesses about race as well as actual numerical errors, all supporting the long-standing official belief of the time, which was that Western Europeans were the smartest people of all.

Never mind that Morton's ancient Egyptian skull collection not only showed no size difference between what he called Caucasian and black skulls, but also showed that both sets of skulls were smaller than those of their African contemporaries. To his dying day Gould didn't believe that Morton was intentionally misleading people—an opinion that rested in the fact that Morton had done nothing to cover the tracks of his methodology. Instead Gould supposed a long-established belief in racial ranking "so powerful that it directed his tabulations along preestablished lines." Keep in mind that Morton was considered among the most objective scientists of his time, the man whom even recognized scholars counted on to "rescue American science from the mire of unsupported speculation." Once again that lofty goal of scientific objectivity, long heralded, had become severely hobbled.

Standing in part on Morton's work, a 1912 *Psychological Bulletin* article by Columbia graduate Frank Bruner declared that blacks were, among other things, "improvident, extravagant, lazy, lacking in persistence and initiative and unwilling to work continuously at details"—and further, showing "a woeful lack of power of sustained activity and constructive conduct." Two years later the manual that accompanied the Stanford-Binet intelligence test—a test still used today—bluntly stated that the enormous racial differences in intelligence "could not be remedied by education." Likewise IQ tests, which have long claimed to be totally objective and unbiased, have been anything but. In one set of IQ tests given to immigrants, eugenicist Henry H. Goddard determined that 83 percent of Jews, 80 percent of Hungarians, 79 percent of Italians, and 87 percent of Russians were feebleminded. Such pronouncements fueled tragic consequences, including people being deported, institutionalized, and even sterilized by their governments.

It was the overwhelming acceptance of these hallucinations—applied not just to people of color but to Southern and Eastern Europeans as well—that in 1924 led to quotas being placed on people immigrating to the United States. Such limits were cited as a necessary step to keep from lowering the national intelligence. As a bit of a reality check, it's notable that since 2000, 40 percent of American Nobel Prize winners in physics, chemistry, and medicine

have been immigrants. Nonetheless, after that 1924 decision, equally reckless claims continued to be made about
Jewish people, one day to feed one of the greatest failures of
human intelligence in history—the so-called Aryan science
of the Nazis.

Today, a growing number of studies show physical health
being compromised in those on the receiving end of racism.
Racist encounters, or even the fear of them, have the effect
of kicking up stress hormones like cortisol—resulting in
chemical reactions that compromise the body's ability to
regulate the heart and the immune and neuroendocrine systems. In a recent study led by Dr. Amani Nuru-Jeter at the
University of California, black women on the receiving end
of racism were found to suffer chronic low-grade inflam-
mation, which puts them at greater risk for serious illnesses
like diabetes and heart disease. It's undeniable that leaving
entire groups out of the mix doesn't just deprive us of
their skills and perspectives. It also threatens our collective
well-being.

I was twenty-one when I left my home in northern Indiana,
making a beeline for the Rockies like a parched man running for a well. Which was exactly what I'd told my parents
I was going to do some eight years earlier, at thirteen, showing up one summer evening in the living room carrying a
box of highway maps and 150 dollars from mowing lawns and
shoveling snow, telling them I intended to ride my purple

Sears Stingray to Colorado, some fifteen hundred miles to the west. Like young people often do, I was chafing at my surroundings—the straight roads, the straight lawns and straight furrows of corn, the straight lines of kids waiting outside the schoolhouse door.

When I finally made it to the high country for good, at twenty-one, what I didn't know was that this golden vision of freedom I'd been carrying around all that time was one dripped into me from my Anglo male kin across centuries: this notion of climbing peaks and floating untamed rivers and basking in countless lonely sunsets from the middle of nowhere, the last of the great explorers. It was all enormously compelling.

It would be years before I understood that my lovely, dizzy dreams had taken flight on the wings of my privilege. Not that I was rich. I'd grown up in a tiny house with a front yard the size of a good hotel room, the son of a blue-collar sheet metal worker. But by the dumb luck of birth—being white and male and fairly well educated—it was an easy thing for me to load up my old Pontiac Tempest and head off to higher, more thrilling ground. I was the benefactor of a reward system centuries in the making, one that had been way less generous to women and people of color. I could on nothing but a whim leave a home landscape that, besides seeming a bit boring, had in many places been ravaged by pollution. And in the process I was also leaving behind the people who were suffering most from those conditions, barely giving them a thought as I breezed out onto the plains on my way to setting up a new life in a pristine run of mountainscape.

The older I get, the more I become aware of how, especially early on, my life was alarmingly bereft of diversity. And more to the point, how the size and pervasiveness of my own culture led me to assume that people on the whole moved on the same playing field as I did. Living according to what nature has to teach has for me meant unearthing something I first heard about as a teenager from a group of women in the environmental movement: namely, that the oppression of one group, one culture, one gender, one place, is the oppression of all.

Back in 1620 Francis Bacon cautioned about the danger of moving too quickly when it comes to declaring universal truths, stating that such temptation was an evil "against which we ought even now to prepare." But science itself was guilty for far too long of leaning on the imagined "universal truth" that the secret to getting back to the Garden of Eden wasn't a matter of deepening our relationship with the Earth, but of learning to control it. And not surprisingly, those deemed best suited to control it were those already in control.

Excluding large portions of humanity from guiding the culture, while at the same time fiercely exploiting nature for its riches, we entered what the ancient Greek storytellers called "hamartia." Hamartia refers to a fatal flaw in otherwise heroic characters—something that compromises them, no matter their good intentions, causing them in the end to make wrong assumptions that then lead to their missing the mark. What starts out as a promising path can through hamartia become tragic, a disaster.

This is a fundamental challenge of being human—a mainstay of the novels we read and the movies we see, popping up almost daily in news about celebrities and politicians seeking power and control only to be brought low. They are not unlike Victor, the lead character in Mary Shelley's famous novel *Frankenstein,* who wants badly to become a famous scientist. The fire of his ambition turns to arrogance, causing him to create a monster that in the end leads to his own downfall.

One of the antidotes to hamartia is the wisdom of nature. As the Oglala Lakota chief Luther Standing Bear put it: "The old Lakota was wise. He knew that man's heart, away from nature, becomes hard; he knew that lack of respect for growing, living things soon led to lack of respect for humans too." Taking our clues from nature about the essential value and strength that arise out of diversity, it's time we get on with looking for ways to better anchor it in our daily lives.

Admittedly, this particular master lesson of nature—the fact that the more kinds of things a system has, the more it thrives—can be a tricky one to embrace. We may feel ready to reach out and connect with other people and other things, but then fall into the trap of thinking we're connecting with others when what we're really doing is trying to get them to be more like us.

In the nineteenth century, lots of people had genuine compassion for what they saw as terrible conditions on American Indian reservations. Out of that concern a solution was hatched based on what became the infamous Indian boarding schools. These schools were designed to "liberate" native people from all non-European aspects of their traditions—from language to clothing to songs to ritual. The idea, much applauded at the time, was to extract native children—with force if need be—from their rooted identity and remake them in the image of a superior white culture. "Trust us," was the sentiment, "we know what's best for you." In the end the Indian boarding schools resulted in thousands of children losing precious threads of identity from their cultural and family ties; many would even die from disease, malnutrition, and physical abuse.

Looking back on such mistakes from where we are now can prompt feelings of guilt and shame. Of course it does. But getting stuck in that shame is yet another kind of self-indulgence, a fruit available only to the privileged, another kind of hamartia that will make us once again miss the mark. On the far side of our regrets lies the question of what we can learn and how we can move beyond learning to act in wiser ways in the here and now.

There's a two-thousand-year-old Chinese story that can help keep us grounded in that task. One day, the story goes, an exhausted seabird, blown off course by heavy winds, landed outside the capital city of Lu. The royal marquis, a man with a very good heart, was much moved by the poor bird's plight. He ordered the bird to be taken to the temple, where he gave it silver bowls filled with his best wine. He then called for a bull to be slaughtered for a royal banquet, that the bird might dine on the very best delicacies of the kingdom. The marquis even summoned the best musicians in the land, playing the songs of heaven, to amuse the bird. Alas, the bird remained timid and confused. By the third day it was dead.

The marquis never realized that the bird had its own nature, its own requirements for well-being. Had he understood this he might have returned it to a protected roost near the coast, left morsels of fish for it, allowed it to be comforted by the presence of other seabirds. The ruler's mistake, his "hamartia moment," came from the fact that rather than taking the time to understand the true nature of the bird, he assumed that in the throes of its trauma it would want exactly what the ruler himself would want in that situation. His thinking—and this is still a common mistake for so many of us—was to assume his own tastes and preferences were the logical standard for all. And as that old Chinese story posits in the end, if everyone followed their opinionated mind as the standard, their actions would inevitably leave out everyone else. Who among us does not have such a standard?

# Healing the Planet, and Ourselves, Means Recovering the Feminine

The feminine values are the fountain of bliss. Know the masculine, keep to the feminine.

—Lao-tzu

Suppose we were lucky enough to drop in on the spectacular sprawling landscapes of Kenya, free there to roam among the baobab and she-oak and blood lilies and sycamore figs. In time—maybe today or maybe tomorrow—we'd surely come upon herds of that most captivating of large mammals, the elephant. Having been around for some five million years, the elephant is widely acknowledged to be among the most intelligent and adept species on the planet. These five or six tons of sturdy yet nimble and playful beings hold the largest, most intricately folded brain of any land creature on Earth.

As we settled down in, say, the semideserts of Tsavo

National Park to observe a herd of forty elephants around a watering hole, we'd likely notice one older female in particular, the matriarch of the herd. She'd be constantly wandering either through or around the outer edges of her group, keeping a careful watch on them, calming and nuzzling her companions while all the time listening and sniffing the air for signs of danger. At just over fifty years old, she carries vast storehouses of information critical to the well-being of her herd. She's an expert not only at mounting defenses against the lions that frequently show up here, but also at using the complex mental maps she holds to lead the group to food or water even when it's twenty or thirty miles away, including watering holes she may not have visited for decades.

That knowledge alone can mean the difference between life and death. Research has helped us learn that elephant groups living in wildlife preserves led by younger females will in times of severe drought tend to stay put, which can lead to a greater loss of calves. By contrast, this group, led by the older matriarch, is more likely to migrate out of the park during big dry spells, moving into faraway places where there's likely water to be found. Thus ensuring the survival of everyone, young and old alike. Unfortunately, this fact also reveals an especially insidious aspect of ivory poaching. By focusing on the largest, oldest members of a herd like this matriarch, which are the animals that carry the largest ivory tusks, poachers often end up ripping away a critical source of wisdom for the entire group.

The fact that this elephant is such a remarkable source of

maternal wisdom is in large part because, as with most elephant matriarchs, she was the eldest daughter of the previous female leader. Starting when her offspring was quite young, this elephant's mother put a lot of attention into carefully grooming her daughter for the job, imparting over many years the knowledge she'd need when her own time came to lead the herd.

And that wisdom she now holds goes well beyond just knowing what to do in dry spells. Along with many of the other older females, she devotes considerable time to those elephants coming into estrus. This elder group will also attend the births of the babies, and then, with help from younger, pre-reproductive elephants, everyone will pitch in to help care for the newborns. This allows new mothers to take more time feeding themselves, which means they can better produce the breast milk their babies need to live.

The prized quality in elephant leaders like this one at Tsavo isn't dominance, as it can sometimes be with other mammals—though clearly, few would want to cross her. Rather, it's her ability to solve problems or, better yet, prevent them. If we stayed around long enough we'd see that she's enormously skilled at brokering agreements, allowing those elephants who have differing views and personalities—and these typically range from gentle to playful to aggressive—to exist in peace for the good of the group. And as happened with her, when she passes away, one of her daughters—or in some cases, if the herd splits, two of her daughters—will step in to lead.

Elsewhere in Tsavo today, just a few miles away from that magnificent herd of elephants, are a couple of lion prides. In these families, too, certain females are essential to survival. Watching them from a safe distance, in the grassy swales that reach beyond a ragged line of baobab trees, we can see that the lionesses are the ones doing most of the hunting, and in stunningly cooperative fashion—either by circling prey or flushing prey toward other lions waiting nearby. If we were lucky enough to come back when there were new lion cubs in the family, we'd find the playful commotion of so-called nursery groups, with many lions, females and males alike, pitching in to help raise and defend the young. Like some other mammals, mature female lions actually synchronize their estrus cycles, which means they can all give birth at roughly the same time; and that makes it easier to take full advantage of this brilliant form of extended child care.

Leaving the elephants and lions of Tsavo, we might head west across Tanzania, skirting the southern shore of massive Lake Victoria, there passing under shadows cast by the seven-foot wingspans of soaring fish eagles, finally pushing into the jungles of the Democratic Republic of the Congo. There we find a close relative of ours, the clever bonobo— an animal with whom we share some 99 percent of our DNA. Like the elephants, the bonobos are being led by matriarchs. Older females hold the highest social ranking in the group. They, and no one else, decide when to move and where. If during our visit a bonobo of either gender managed to take down a forest antelope and then showed it to the group, an elder female would walk over and put her arm over it. Only then would we see everyone else come around and hold their arms out, waiting to get their share. And at any time during our visit, should an aggressive male start harassing a young female, maybe trying to have sex with her against her will, let's just say he can expect a tree-full of trouble from the mature females.

In countless species, from meerkats to whales, females hold both nurturing and leadership roles. In orca society, the bonding of females to their offspring is so strong that an adult male is eight times more likely to die within a year after his mother's death than when she's alive. In short, the essence of feminine mammalian energy goes far beyond re-production and includes the vital, complicated tending of relationships critical to a thriving life. For those who look deeply into the woods, the desert, shoreline, or mountains, the jungles, oceans, or prairies, it fast becomes apparent that there's simply no other way for life to survive.

The fact is, in mammal species where males and females are roughly the same size, female leadership is often the norm. Even when males are larger, such as in chimpanzees, gorillas, lions, and wolves, it's often still females who make critical leadership decisions. Yes, female leaders in the ani-mal world can have impressive physical strength, but one of the reasons they lead is because of the vital skills they have for building coalitions. From hyenas to elephants to hu-mans, female wisdom and relational instinct, way more than strength alone, are a critical part of survival.

This isn't to take anything away from the essential role of masculine energy in nature. The young males of the Tsavo elephant herd are staying with their natal group ten or even fifteen years before heading off on their own, being full participants in helping to bond and defend the group. After they reach breeding age and have left their natal group, sometime in their twenties, they'll follow strict social rules for copulating with willing females—for instance, first gently laying their trunks across the back of a female in estrus to see if she's receptive.

Meanwhile in the wolf world, and in lion society too, males play critical roles in both hunting and defending territory. But also in caring for the young—both by engaging in endless rounds of play, but also by going out and securing food for the cubs. Again, nature has created a world where the success of elephants and wolves and lions and countless other species comes from a full expression of both sexes. The idea that one gender is more important than another is a human illusion—one that ignores the fact that nature is an expression of the balance between the two. Plain and simple, life thrives when the masculine and feminine are fully partnered.

Even so, for much of the past four thousand years humans have been hard pressed to live with that truth, choosing instead to move through the world with masculine energy as their chief guide. Arguably, of all the wrong turns we've made when it comes to living well with the Earth, living sustainably, it's hard to imagine a more calamitous misstep than this.

So how did we end up in such a strange place—ostensibly struggling to understand the world, all the while having blinded ourselves to fully half of its wisdom? Actually, there was a time when humans were a lot more inclined to celebrate and nourish the essential skills and insights of the feminine. We know this in part from looking at the myths of ancient cultures. Tales pointing not so much to flesh-and-blood girls and women, but to a much bigger, more archetypal feminine energy. We know from these stories that such energy was seen as essential to life itself. Beyond its obvious role in bringing new life to pass, feminine energy also was a doorway to sustaining it. For the humans sharing these stories with one another, such energy helped nourish their search for well-being, guiding them to the power of creative relationship. Benefits, by the way, available to men and women alike. Stories rich with both masculine and feminine energies served as reminders of the integrated nature of the world—of the fundamental truth that life sparks, sprouts, and then flourishes in a commons of interconnection.

Some of the old tales that highlighted feminine energies had to do with the creation of Earth. Like this five-thousand-year-old myth from the ancient Pelasgians, who lived along the fertile, sun-drenched coast of the Aegean Sea:

In the beginning there was no order, no stability—only chaos. But then came the time when from inside that chaos there arose a great and powerful Mother Goddess. Seeing the jumbled state of affairs in the universe, she got to work— first separating the water from the sky. When that was done the Mother Goddess stood back and took it all in, and saw that it was good. In fact, so pleased was she that she placed her feet on the water and began to dance. And she kept dancing, too, moving toward the south, surging and twisting with the waves of the sea, becoming so filled with slip and sway that a mighty whirl of wind began blowing in her wake, like a small tornado. Mother Goddess turned and caught that powerful wind, caught it in her hands, rubbing and rolling it between her palms until a great male snake appeared.

A quick aside here, in case you may be shuddering at the mere thought of snakes: As far back in history as we're able to see, often puzzling together stories from eroded marks on long-buried stones, the feminine has been linked to the snake. For example, the people of ancient Macedonia and Crete were led by Great Goddess figures who routinely kept the company of snakes. In part, this is because of the snake's uncanny ability to shed its skin, which led to people's associating it with birth and fertility. What's more, snakes live on and even below the ground, which means they're always close to the life-giving powers of the Earth. And finally, the venom of some snakes was said to hold properties similar to those of certain plants and mushrooms—ingredients employed in the feminine-driven arenas of physical healing and spiritual transformation.

These associations are so strong that even today we still use as our chosen symbol for doctors and hospitals the so-called staff of Asclepius—two snakes twined around a center pole with a pair of wings at the top.

But back to our story.

Mesmerized by the goddess, the snake came to curl himself around her seven times, clinging fast as she danced. Afterward the goddess assumed the shape of a dove riding the waves of that primal ocean, and when the time was right she laid what was known as the universal egg. The snake embraced the egg until it cracked in two and hatched. When it did, out tumbled the goddess's children: the moon, the stars, and the planets, including Earth, with all its mountains and rivers and trees and flowers and all manner of living creatures, including humans.

The energy represented by the snake could land in the world only through the feminine energies of the goddess. The creative power of the goddess, on the other hand, reached its highest expression when ignited by the male energy of the snake.

Similar stories from the ancient world abound. In Asia, the Balkans, Greece, and North Africa were endless streams of stories about Cybele—yet another manifestation of the Great Mother, sometimes called Mother of the Gods. Like other feminine powers, she was credited not just with birth but with granting healthy, thriving life as well, employing a miraculous ability to heal sickness in creatures great and small. Being exposed to the notion of feminine energy—in the stories you listened to, in the sacred objects you saw, and in the prayers you prayed—would go a long way to help shape your perspective of the world.

You might have imagined seeing these feminine powers of nourishment and healing in the fennel and saffron you collected to make dinner, or in the elderberry you gathered to treat the burn your son got the day before while helping tend the cooking fire.

You likely would be imagining as well the feminine in the larger *processes*, or cycles of nature. The daily comings and goings of birds. The rising of the sun and the passing of the stars and the moon overhead in the black of night. The spin of the seasons—summer turning into fall and then winter and then spring, and finally summer again.

Today we can look into wolf and elephant and beaver and lion societies, and hundreds of other animal cultures too, and still see the "agency" of masculine energy—used by males and females alike—to accomplish specific tasks like hunting, or establishing and defending territory. At the same time, in each of these animals would be feminine "relational" qualities—again, available to both males and females—used to keep the peace within the herd or pack or pride, to care for the young, to strengthen bonds. All of which is essential to survival.

The same balance of masculine and feminine that nourishes these animal cultures can nourish us, as well. We, after all, are nature too. What's more, we have the kinds of brains that can note what works in the world around us, and then choose to make it more prominent in our own lives.

Even today, certain stories from those other times and places can serve as poetic reminders to us about how the bigger world really works. Take this tale from ancient Samaria, known as "The Birth of Wood and Reed."

> In the beginning, there was only the sea. And the sea was called "Mankind's Mother." In time this vast primordial sea gave birth to An, who was also known as "Heaven," as well as to Ki, who in our world might best be thought of as the Earth Goddess. One bright morning the Earth Goddess was overcome with delight at the beauty of the world before her. Flushed with inspiration, she began adorning herself in a flowing weave of the finest greenery, a brilliant shawl of flowers and leafy vegetation to which she added stunning ornaments of diamonds and diorite and silver. An,

meanwhile, who happened to be looking on from his heavenly realm, found himself enormously attracted to the vision that was Ki—fairly swooning before her blooms and stones and the emerald green of her pastures.

So An had an inspiration of his own. He put on his most regal clothing, and thus festooned descended from the heavens into one of Ki's beautiful meadows. And there Heaven and the Earth Goddess quickly fell into an adoring embrace. On that mystic day they made love, unfolding a dance of the utmost tenderness and passion. As a result of their embrace, Heaven deposited the seed of Wood and Reed into Ki's womb of soil. And with that a brand-new flush of life was set to rise—this time in the form of elegant trees and sinewy reeds—all soon to flourish from the body of the Earth Goddess.

As we later learned from science, every living being does in fact owe its existence to this union of Earth and Sky. The sky pours sunlight onto a fertile planet—the essential ingredient that allows plants to sprout, grow, and set seed. And those plants, in turn, are the very foundation of what sustains the lives of the animals, birds, and insects. The story of Ki and An is something you might easily dream up all on your own some warm spring afternoon, out strolling in a meadow bedecked with wildflowers, crafting the tale from nothing more than having taken a long, deep look at what's going on around you. What may be missing from your first take, though, yet worth considering, is the notion of that meadow as proof of the astonishing power of an interconnected masculine and feminine.

It's true that today such holistic thinking still feels new, even strange. Only twenty years ago our scientific research into climate change was focused almost exclusively on what was going on in the atmosphere overhead, in particular, figuring out how that atmosphere behaves when it gets loaded with carbon dioxide.

Today, though, researchers are beginning to quantify the spectacular amounts of moisture in the atmosphere being generated by the world's forests—forests that in some cases are quite literally creating what amounts to rivers in the sky. So, yes, increased carbon dioxide can cause drought by warming the atmosphere, proving regionally disastrous to growing crops. But now we also know that down here on the ground, failing to protect a certain big forest could mean stealing the rain from those same farms, even if the forest and farms are thousands of miles apart.

In a very real sense, then, we're tying the earth and sky together again, not unlike what happened when Ki and An got together in "The Birth of Wood and Reed." To the extent scientists continue to uncover the striking unity of things, linking sky and earth and forests and soil and plants and insects, they're creating fresh opportunities for us to once again embrace that feminine part of the equation that speaks to relationship. Science, of course, uses very different language and images than myth. But in the end it can nonetheless spur us toward realizing that all of nature, which includes all of humanity, is served best when masculine and feminine energies are honored as equals.

Joseph Campbell used the term "the great reversal" to describe that time when the holiness of our own nature and the nature of the universe slid away under canons of belief that imagined us as captives to individual bodies and to the Earth, locked in an insufferable state of sin. A big reason that such an idea could ever spark and burn is due in part to our having turned our backs on the feminine.

While Campbell was referring to a time beginning around 600 BC, the foundation for that shift was arguably laid some three thousand years earlier, in a time when agriculture became concentrated, allowing ruling landowners greater wealth. Before long, as if a contagion, tyrannical leaders were launching war after war in attempts to grab that wealth for their own kingdoms. It was in the midst of all this bluster that the very mythical fabric of cultures began to change. Throughout the many centuries it would take before Campbell's great reversal was complete, powerful rulers were quashing nature-goddess stories in favor of solo acts by blustery male gods.

The storylines of this new emerging canon went something like this: First, a great goddess figure would be overpowered or killed by a god. Then the male god took possession of what were formerly feminine energies, feminine responsibilities. As the centuries unfolded, the motif evolved into something more aggressive still, as the male god merged into a powerful so-called storm god. From that came the era of the full-blown male creator god—a supreme ruler capable of creating the Earth with no voluntary contributions from the feminine. Instead of exercising power of their own, the great nature goddesses were, at best, relegated to intermediaries tasked with appealing to a male god for mercy on behalf of suffering humans. Far more than a meaningless shift away from goddess narratives, this massive reversal in worldview marked the beginning of forgetting those common threads we share with everything on Earth. Achievement, it turns out, needs wisdom. And without the feminine, wisdom was flying about with clipped wings.

As the stature of the goddess and the Earth she nourished began to shrink, so, too, did the lives of real women. In the older, goddess-rich cultures of ancient Sumer, women ran businesses, became physicians and scribes and priestesses, were judges in court. By the time of ancient Greece, most could no longer own property, couldn't divorce, vote, or hold public office.

Standing against a growing insistence by patriarchal leaders to be in control, the feminine energies—and all the creative improvisation they represented—were derided and suppressed as a threat to masculine power. The Greeks started spinning tales like the one about Apollo going after the Furies, who'd begun their mythical lives eons earlier as an ancient group of virgin deities said to represent female power. The Furies had long been revered for dispensing justice, especially on behalf of nature. The story tells us that this was especially irritating to Apollo, who'd specifically gifted mortals with the power to bring nature to its knees. In the end, Apollo defeats the Furies by arguing against them in a court of male judges, winning a pardon for a god named Orestes who'd slain his own mother.

The Furies wanted Orestes punished. But Apollo argued to the judges that it was the male who was the active agent in childbirth, the woman being nothing but a vessel. He argued to the court that the lack of an actual blood relationship between a mother and son meant that any son killing his mother was no worse than any man killing a stranger. The court agreed. The inability of humankind to control nature, represented by the Furies, the shadowy commotion from which creation itself was said to spring, was now seen as a vanquished foe.

Animals previously associated with the feminine, like the snake, were made evil, which is how the ancient Hebrews came to cast the snake as an emissary of Satan, a creature later woven into a story meant to prove women untrustworthy: the story of Eve. By AD 200, the so-called father of Western theology, the Roman Tertullian, declared flat out that women were "the doorway of the devil."

It took centuries to happen, but feminine qualities, including the rich interconnectedness of life they and their companions represented, drifted out of reach.

This illusion—that the world turns not on an equal partnership between masculine and feminine, but by the masculine suppressing and controlling the feminine—has pretty much been with us ever since. In the 1800s, Europe and the United States delivered a flurry of scientific findings "proving" that women were incapable of having original ideas. If females wished to participate in science, it could only be to shed new light on the theories of men, a rule promoted by even the most progressive male scholars of the day. By the early twentieth century, women were being routinely warned away from higher education, cautioned that it would lead to defects in their sexual organs.

In other words, by suppressing women's voices even science ended up throwing a lot of shade onto the feminine. And given the influence that both science and religion have in the culture at large, it's hardly surprising that such bias would spill out into every corner of life.

It certainly washed over my own life. My boyhood, like that of others, came wrapped in hundreds of such messages big and small—from notions that sports were best left to boys, to the lone-male storylines in the Westerns I watched on television, to the near absence of women in my schoolbooks.

Even the nature books I gobbled up as a kid were slanted

toward male-centrism. In fact between 1900 and 2000, of the most popular children's books set in the natural world, male animals were almost four times as likely to be cast in the lead role as females. No wonder that when I went to the zoo I heard my parents, as well as other kids—boys and girls alike—more often than not referring to the more captivating animals as "he."

Such are the stories we've been telling ourselves. And as anthropologist Peggy Reeves Sanday noted, held within the myths and stories a culture tells is "a basic statement about their relationship with nature and their perception of power in the universe."

Little wonder that my friends and I ended up with notions of actual girls and women being less important, less capable. At the same time, such bias made us strangers to the *archetypal* feminine—to those qualities the goddesses had held up for all to see, males and females alike, including creativity and relationship and nurturing.

It was only in my later teens, when I was really steeped in the natural world, actually seeing how these energies were entwined, that some of these earlier assumptions began to unravel, including, happily, many of the misguided conclusions I'd made about the value of the feminine in boys. In men. In me.

Today we stand without a single enduring creative or scientific tradition or institution, not art or music or writing, nor philosophy or psychology or medicine, that doesn't still express remnants of this lopsided, masculine-dominated trajectory.

Even history and anthropology still struggle with this limited vision. The prevailing hunter-gatherer theory of early human behavior has long seen men in the role of hunters, supplying the bulk of the food, while women care for the young and gather local edible plants. Largely absent from the model is any nod to the roles that women—like so many other female mammals—may have had in directing and sustaining alliances or prompting peace deals. At this point, we really don't even know how much actual sustenance came from hunted animals versus the plants harvested by women. It's like assuming that just because the male leader of a wolf pack is physically stronger, that his actions are the critical keys to delivering elk to the pack. Yet beyond the fact that female wolf leaders can be also powerful hunters, as often as not they're the ones who determine where and how to find the elk in the first place.

Relegating the feminine to the back seat has actually changed the way we experience reality. Again, the feminine in nature was long credited in mythology as being the force that unifies, that binds in relationship the human with the deer, the deer with the grass, the grass with the sun, the sun with the rain, the rain with the ocean, and the ocean with the swelling and waning of the moon—reliable, yet at the same time often spectacularly creative and improvisational.

Going through life stripped of that sense of unification, of relationship, we came to see a person, a tree, an animal as stand-alone beings, which has often caused us to miss the crucial connections they hold. We might wipe out sparrows because they're eating our rice, as China did in 1958—a masculine-oriented "action" step—only later to realize that in the absence of sparrows, insects will proliferate, devouring rice and leaving us with far less of it than we had before.

To see through a feminine lens, suggests the ancient Tao-te ching, is to see that which "clothes and feeds all things but does not claim to be master over them." Rediscovering this quality of inclusiveness, learning to "hold it all," allows us to move past the idea of saving just small pieces of the planet and to focus instead on saving the whole system.

When we use nature as a guide, what we discover is that there's nothing more fundamental to a healthy and resilient life, both inner and outer, than restoring the feminine—literally and archetypally—to its rightful place in full partnership with the masculine.

Right this minute, out in the hinterlands, hundreds of alpha female wolves are doing what they do best, making choices about where to travel to find prey, disciplining underlings, fighting in territorial skirmishes, and choosing safe summer rendezvous sites from which to care for their young. And this year or next, in the face of drought, elephant matriarchs will lead their herds with uncanny precision to distant watering holes, eyes and trunk tuned to any dangers that might be lurking in the bush. The world goes on and on in just that way, with female leadership showing itself in everything natural from lions to bees, lemurs to orcas, antelope to octopus to hyenas to bonobos, prairie chickens to Harris hawks to kingfishers. Standing not under or over the masculine but equal to it, bringing the critical balance on which the survival of the species absolutely depends.

LESSON FIVE

# Our Animal Cousins Make Us Happier—and Smarter

[Animals] are not brethren, they are not underlings;
they are other Nations, caught with ourselves in the
net of life and time.

—Henry Beston

Known only as 14, she was one of fourteen wolves re-introduced into Yellowstone National Park in 1995; the youngest, and the least well-known, of what wildlife biologists would years later call the heroine wolves of Yellowstone. Her life was a mystery, in large part because she, her partner, and the pack they so faithfully guided, the Delta pack, were destined to live in one of the most remote reaches of wilderness in the continental United States: the southeast corner of the national park, in the magnificently untrammeled twenty-mile-long valley known as the Thorofare.

Shortly after her release, following six weeks spent in an

123

acre-size acclimation pen, 14's life took a curious turn. On a routine tracking flight, biologists discovered she was denning, about to give birth to pups north of Yellowstone Park near the small village of Roscoe, Montana. Smack in the middle of the Lazy E-L cattle ranch. Her pack included two other adults and her mate—an aged grayed wolf known as Number 13, whom privately the biologists had given the nickname "Old Blue." They were model citizens, never once preying on livestock despite being surrounded by cattle every day. Several of the ranch cowboys actually grew a little fond of her, especially, one of them telling me it was "kind of nice" to see her out there roaming about with the elk.

But some of the neighbors thought otherwise. After wildlife officials received numerous death threats against 14 and her pack, they decided to launch a capture effort, hoping to relocate the original adults and four young pups to a new home on less contentious ground.

The relocation a success, the eight-member family thrived for the next year in the southeast corner of Yellowstone. This is a challenging place for any animal, covered for much of the year by deep snow—a place where resident elk herds can in the cold months be hard to find, most of them drifting away in late fall onto more suitable winter range. Number 14 and Old Blue, though, bonded for life, proved brilliant at leading their pack over yawning distances to find those hidden elk. And much of that leadership fell to Number 14.

Soon, by the summer of 1997, Old Blue was showing his

age. More and more frequently biologists on tracking flights saw him struggling behind the pack. He was unusually slow to join in when it came time to hunt. During close-up contact the biologists had with Old Blue during winter collaring efforts, they noticed his teeth badly worn—a sure sign of advanced age. As a result, other adults would regularly step in to open the tough hide of a recently killed elk, and then, acknowledging his status, back away to allow Old Blue to eat first.

Then in January, Old Blue's collar began emitting the fast, steady signal known to biologists as mortality mode—which happens when a wolf is no longer moving. The scientists guessed, and they were right, that Old Blue was gone.

Following the death of her mate, in a move no wolf researcher had ever seen or heard of, 14 took off. She left her home territory near Heart Lake without her pups and yearlings, which to the biologists seemed extraordinary behavior. She was tracked by plane as she wandered west through deep snow, crossing high terrain so inhospitable it contained not a single footprint of another animal. Through a combination of tracking her from the air and actually following her prints on foot, the wolf team would finally locate 14 on the outer reaches of the Pitchstone Plateau. There she stood alone, ten thousand feet high on an empty, windblown slope, peering at their research plane circling overhead. Then she resumed her journey, traveling on for another fifteen miles.

A week later, she returned to her own territory and re-united with her family. Though no one wanted to say 14 traveled alone because she was mourning the loss of her mate, one biologist quietly admitted to me over a beer that he'd been wondering if that was exactly the case. That it all came down to a matter of deep grief. To this day he hasn't gotten that thought out of his head.

There are few things more deeply stirring than the recognition that we share the world with countless other creatures that, across their own measure of days, yearn and play and feel anger and fear. And grieve. It's a sense of communion that fires the heart, which is why so many stories from around the world celebrate life in just this way, routinely blurring the lines between humans and wolves and eagles and rabbits and bears and whales.

That this commonality with animals could ever become suspect says much about how in more modern times we learned to see the world. A perspective that for a good num-ber of centuries had no room for the thought that we hu-mans, supposedly so vastly superior, would share thoughts or emotions with the finned and feathered and four-leggeds of the world. Now, and increasingly, we know better. As evolutionary biologist Marc Bekoff notes, there are a great many mammals and even birds that have brain systems and chemistries similar to those linked to human emotions. Do-pamine levels increase in smitten, besotted rabbits no less than they do in infatuated humans. Plenty of mammals during times of attraction experience internal shots of the hormone oxytocin from their pituitary glands, which for all of us is strongly associated with courtship. Recent scientific findings taken together, Bekoff notes, suggest that at least some animals are capable not just of grief and happiness, but also of love.

To be fair, that wolf biologist's reticence to say much about grief in Number 14 is part of a strong and entirely understandable resistance scientists have to so-called anthropomorphism. Literally translated as "human form," the term has to do with projecting human qualities onto other species—imagining them to be sad or happy or any of a wide range of other emotions that we ourselves feel on a regular basis.

And still. I recall in the early 1990s sitting in a wildflower meadow at the foot of Montana's Gallatin Mountains with a Northern Cheyenne elder. At one point I decided to ask him about the idea of anthropomorphism. When I did, he looked at me sideways and shook his head.

"To us, anthropomorphism makes no sense," he said. "Can't you see? We're the ones who took our qualities from the animals. It's never been the other way around."

The peace and contentment we manage on this hard journey through life has much to do with being able to acknowledge what's precious, worthy of respect. Not all of us had childhoods like that described by the Northern Cheyenne elder, where adults carefully, patiently instilled in him a deep, reliable sense of the world's interconnectedness. But even for the millions of us who didn't get that particular teaching, animals can help us find our way. With their ability to enchant us, animals have the uncanny power to leave us always wanting to know more. We're drawn in to what they do and how they do it—eager to step toward these other lives, wanting the world to be big enough and safe enough to hold us all.

Animal expressions of bonding—the sheer joy they display, along with the resulting grief when those bonds are broken—are hardly limited to wolves. In fact, we've seen them in everything from geese to ducks, horses to rabbits, cats to whales to sea lions to chimpanzees. In 2004 Tina, a beloved Asian pachyderm living at the Elephant Sanctuary in central Tennessee, passed away at the age of thirty-four. Immediately upon her death three of her friends entered her stall and began touching her body with their trunks—a behavior recorded on many occasions with elephants in the wild. Later, the day after sanctuary veterinarians had performed a necropsy on Tina and then carefully buried her body, the other elephants abandoned their normal spacing behavior to huddle together shoulder to shoulder, looking at the place where she was laid to rest. Tina's best friend at the refuge, Sissy, took her own favorite toy—a car tire she was enormously fond of—carried it over, and left it on the grave.

Similarly, a few years ago I was writing a biography of Carole Noon, a remarkable protégée of Jane Goodall. Noon founded the largest chimpanzee rescue sanctuary in the world—saving more than three hundred chimps, many from devastating lives of neglect and abuse locked in concrete-floored cages at various research labs around the country. One chimp named Phyllis, living in New Mexico, had been part of the original US space program, born to a mother captured in the late 1950s in Africa and brought to live in a lab in at Holloman Air Force Base. There Phyllis was prepared for possible space flight, including spending time in what was known as "couch training." For this she was placed in a suit covering her entire torso, which was then tightly lashed to a horizontal metal frame to keep her from escaping. The experience was so unpleasant for Phyllis that she ended up tearing deep lesions in her legs trying to escape. With that, the decision was made to pull her from the training.

But that didn't end her suffering. The following year she was moved into hepatitis research, knocked out dozens of times so veterinarians could perform biopsies on her liver. After that she was placed into a breeding program, shuffled from male to male for as long as needed for her to become pregnant. Otherwise she was alone in her cage. The babies she gave birth to were taken away hours, or at most days, after birth, exactly as had happened to her. Not surprisingly, her medical records during that time list a litany of symptoms indicating severe depression. Finally, when Phyllis was just past her thirty-second birthday, Carole Noon came along and rescued her, taking her to a life in Florida with grass and good food and toys and open sky. And, most important, other chimps with whom she could make friends.

Carole Noon was thoroughly beloved by those chimps. She often spent hours in the middle of the night sitting and talking with them, comforting them when they seemed afraid or anxious. When Carole passed away—dying from cancer in her home beside the refuge—the chimpanzees emerged from their shelter buildings to stand together in silence, solemnly facing the house. Paying tribute. Overcome with sadness at the passing of their champion and friend.

The year 2018 saw another heart-wrenching display of grief capture the world's attention, when an orca named Tahlequah gave birth to a calf in the San Juan Islands of Washington State. It was the first such calf born alive in that region in three years. But sadly, minutes after birth the newborn died. Whales and dolphins, like elephants and chimpanzees, are well known for grieving the loss of friends and family members—rituals that typically last a day or two. But on the death of Tahlequah's calf she began an extraordinary grieving ritual, swimming north for a thousand miles over seventeen days, the entire time pushing along the body of her lifeless baby, diving down through the cold blue waters of the eastern Pacific to retrieve the body whenever it slid from her nose.

It's probably not all that important, the sense we make of a particular animal's behavior in a given moment. What matters more is that we pause to really feel how deeply we're drawn to them in the first place. As environmentalist Paul Shepard pointed out, anthropomorphism binds our continuity with the rest of the natural world. It generates a desire to identify with animals, to learn about them, "even though it is motivated by a fantasy that they are no different from ourselves." In fact, no animal experiences life exactly like you or I do or like other animals do. While there's likely real overlap, animals also take in the world in ways we can't imagine, different from us as much as from one another. When we really understand this, we gain a golden chance at a new beginning with those with whom we share the planet, a kindling of relationship fueled by mystery and wonder.

Images of newly born bear or lion cubs or gangly moose calves often bring us a delightful buzz in witnessing life beginning all over again. We can *feel* the bond of a mother animal toward her babies. We know deep in our bones about the endless patience she'll have to muster in the face of their persistent need to nurse and play and cling. We can feel the joy of bouncing baby goats. We can easily share in the glorious enthusiasm shown by a dog flying across the lawn to chase down a bouncing tennis ball.

But that warm, fuzzy feeling at seeing the cutest or most playful in the animal world really isn't enough. Our draw to cute needs to rest in a mature understanding of our shared fate. To stop with cute, to remain passive on the larger issue of safeguarding what animals need to maintain their lives, is to abandon the deeper joy and comfort to be found in our intimate connections with other beings. When it comes to other creatures, each and every day brings a chance to find in nature good reasons not to divide and objectify. Rather, to see that the ultimate fate of a bear family, or lion family, or wolf or moose family is, like our own fate, tied to an intricate web of life whose strands are fortified by a healthy planet.

When I was first asked in 1994 to chronicle the return of wolves to Yellowstone, I was elated. The thought of being near such intelligent animals was thrilling—the chance to watch how each would take on the task of making a life in what for them was new, unexplored country. But I was also a little reluctant. While I knew wolves to be beautiful and graceful and clever, they also had a deep mythical sheen. Genuinely feared and at the same time deeply esteemed, wolves had loomed larger than life across the centuries, alternately gods and then devils. The truth of wolves as smart, hardworking carnivores out there trying to make a living like everyone else had too often been lost in either glamour or gloom. And in the mid-1990s, that same tension was fueling a spectacle around the Yellowstone reintroduction so raucous that it caused me to think twice about going anywhere near it.

As the arrival of the wolves neared, circus tents were set up around the northern edge of Yellowstone to handle the hundreds of journalists and film crews from around the world. Meanwhile in Montana's state capital, angry-looking men paraded with signs reading WOLVES ARE THE SADDAM HUSSEIN OF THE ANIMAL WORLD and WOLVES: SMOKE A PACK A DAY. In the blast of a wave of death threats against the wolves, a decoy truck caravan traveled miles apart from the actual transport vehicles, in hopes of foiling anyone who might start shooting. Moreover, so many threats had been leveled against the wolves in the weeks leading up to their arrival in Yellowstone that their acclimation pens were rimmed with high-tech military monitoring devices.

These acclimation pens, where the relocated wolves spent between six and eight weeks, were part of an effort to quiet the animals' instinctual homing response. The location of the park was such that their natural urge to run north toward home in Canada would've taken them into the heart of cattle country—potentially a recipe for disaster. Months later, after the wolves had been successfully released—and after animals like 14 and her pack had been moved off of private ranch lands—the furor and attendant threats quieted down some.

Well after the circus tents had come down, after the journalists covering the wolf reintroduction had headed off to the next big story, a small core group of us stayed on. And we became frequent witnesses to the daily miracle of those fourteen wolves running free—exploring, hunting, setting up new territories in a slice of rich, wild country that hadn't felt the feet of a wolf pack for some seventy years. It was thrilling to watch them in the months and years that followed—from the serenity of the backcountry as well as through chance sightings along the northeast entrance road in the Lamar Valley. It was then, far from myth and fear, that some version of the deep sense of kinship with these creatures that native people had long recognized began to rise in me.

I watched their somewhat madcap, highly communal care for their pups and routinely saw astonishing displays of cooperation around the very dangerous, potentially even deadly business of hunting elk. I lingered to watch the wolves in their times of rest, observing them as they hung out near the den or roamed around, curious, apparently in search of nothing more than what they might find on the other side of the hill.

Over and over they surprised me, left me shaking my head in wonder. Simply the ways they played with one another, and how often, were astonishing. Teenage wolves taking pieces of elk hide, for example, then tossing them into the air and catching them, like so many Frisbees. Or the way entire wolf families, adults and young alike, would turn over onto their backs at the tops of hills to slide down snowfields, then eagerly run up the hill, tongues lolling, and do it all over again.

Any scientist will tell you there's great benefit to keeping distance from the things we study, teasing out conclusions that can then be confirmed by others who also manage an objective perspective. But it was while watching the wolves of Yellowstone that I began to wonder if we might do well to consider that Northern Cheyenne elder's suggestion—that many of the qualities we've developed as individuals and communities have flowed not from us to the animals, but the other way around. Undoubtedly wolves—and many other animals—were our ancestors' teachers. What's more, clever as we may be, we simply haven't outgrown them—nor will we ever—as a source of wisdom. Their perceptual skills, their instincts, their physical strength and grace, as well as their success at forming social bonds, are all things that have been honed over tens of millions of years. From a genetic standpoint all of us Earth dwellers chose different paths. But those who've made it this far, whether dressed in feathers or fur or blue jeans, did so not by luck—though we've all had a little of that—but by astonishing levels of resiliency.

Surely we can find a way to discern biological facts, while at the same time keeping our imaginations open to that which links us but for which we may not yet have words or theories. I actually think this is the resolution the wolf biologist was looking for when he considered Number 14's grief. Along with his impeccable record of scientific rigor, I now know he's also spent a fair amount of time quietly pondering what else might be going on in those animals far beyond his ability to observe or measure. His science, then, is exactly what we need if we're going to meet the challenges ahead in thoughtful, humane fashion, the kind that lives in two worlds at the same time—one filled with direct observation, the other with mystery,

Holding mystery in tandem with direct scientific observation is a big leap from what we saw going on with René Descartes. And Descartes had plenty of company, what with most of his seventeenth-century colleagues also believing animals were without feeling or emotions, bereft of the qualities we most valued in humans. Typically the distressed cries of animals were deemed no more than noise made by a kind of inner spring set to squeal at a hard touch.

A big reason educated people of the seventeenth century had such a hard time entertaining the notion that animals could play and feel and cry and talk, could share important information in their own highly evolved ways, was because such ideas threatened fundamental beliefs about the universe. For science, and for religion too, any suggestion that we were related to animals defied the order of creation.

Yet even early on there were critics of this view. As Michel de Montaigne wrote way back in the 1500s—showing himself to be far ahead of his contemporaries—humans were foolish to set themselves above every other creature.

"When I play with my cat, how do I know that she is not passing time with me rather than I with her?" de Montaigne mused. "Why should it be a defect in the beasts and not in us which stops all communication between us?"

Still, the vast majority of early scientists who studied

animals called proud attention to their faithfulness to that Greek-inspired idea of disconnecting from what they were studying. And yet just as was true with so-called scientific conclusions about women and people of color, deeply held personal and societal beliefs about animals corrupted the researcher's objectivity. Humans were the chosen ones—the ones in charge, put there by God, and therefore free to do as they wanted with the rest of creation. True, the claim that animals didn't feel pain raised eyebrows in some. But all in all it played well into a deeply held societal belief that humans and animals were unrelated. In Greek myth, Apollo had argued for Orestes to be released for killing his mother because she was essentially a stranger to him. Seventeen hundred years later, in the world of humans, men made the case for inflicting all manner of pain on animals because they had neither feeling nor soul.

We really can't overstate the degree to which early science—and through science ultimately society itself—was bent on identifying and then fiercely defending the concept of "the other." This was the road by which entire groups of living beings were labeled as inferior—as "lesser than" the people doing the labeling. When Descartes made this argument, one of his favorite "proofs" of human superiority was that "not a single brute speaks." As if the only communication worthy of being called intelligent is that expressed by human words.

But what about the incredibly complex communications of birds, or the layered conversations and kinship rituals of dolphins and whales? Wolves, coyotes, and foxes can talk to their own kind with more than a dozen facial expressions involving feet and head and body postures, as well as tail positions, along with sounds from barking to growling to whining; all of those get combined into a fairly dazzling number of coherent messages. Then there are the dancing dialogues of bees, used to share the location of flowers with other members back at the hive. A bee's "round dance" passes along information to other members of the hive about resources located within three hundred feet. The more involved "waggle dance," on the other hand, communicates both direction and the amount of energy needed to reach food farther away. Vervet monkeys use one of several kinds of vocalizations to alert kin to specific predators lurking nearby, and they and other primates build bonds by communicating through touch.

We've known for some time that animals, including chimpanzees, use tools, fashioning implements to forage army ants. Dolphins sweep the ocean floor with marine sponges to uncover prey, and ravens make tools from twigs to impale larvae. We've watched crows making toys, and seagulls using cars traveling down the highway to break open shellfish. During egret-nesting season, alligators balance sticks on their snouts, luring in the birds that are rushing about collecting materials for their nests.

Zoologist Alan Rabinowitz, who was an inspired and tireless preservation champion for the world's forty species of wild cats, had a rather different take on the idea of animal communication. Growing up in Brooklyn in the 1950s and '60s, as a boy Rabinowitz was often upset by how otherwise kind people thought nothing of a pet chameleon dying from dehydration, or allowing a turtle to die because they'd let the tank go dry. Or for that matter, even how frequently the sick dogs of friends or family members were euthanized.

"There was a lack of feeling . . . towards animals. I . . . could see that if animals had a voice, people would not treat them the way they do."

As a boy Rabinowitz had grown up with a severe stutter. So severe, in fact, that the New York City school system declared him handicapped and then promptly ushered him into special classes, lasting from kindergarten through the sixth grade. But he found he could do two things without stuttering. One was to sing, which opens up the airflow through the throat and allows the words to keep flowing. The other was to speak to animals. For years he'd leave his school having not uttered a single word and make his way back home, where freely, without stutter, he'd talk to his pets—green turtles and hamsters and chameleons.

In time Alan's father, who could see a kind of relaxed contentment wash over his son in the aftermath of the time he spent with his pets, began taking him to the Bronx Zoo. Specifically they went to what was known as the Great Cat House, filled with lions and jaguars and tigers.

The boy and the big cats traded stares, with Alan often perceiving their sadness, their anger. And always he left the Great Cat House feeling he'd had far more meaningful communication than he'd ever been able to manage with humans. One jaguar he talked with especially often. And what he said to the big cat repeatedly was "[Someday] I'll try to find a place for us."

"I swore to myself . . . that if I was ever able to find my voice, that I would be their voice. I would be there for them."

Which is exactly what he did. In addition to doing groundbreaking research on wild cats around the world, in 1986, at the age of thirty-three, Rabinowitz set up the world's first jaguar sanctuary on the eastern slope of the Maya Mountains, in south-central Belize.

Fortunately, ideas about animals not being able to even feel pain would by the 1700s begin to crumble. Yet the hubris that allowed such notions to sprout in the first place lingered on. Two centuries after Descartes came a brilliant man by the name of Claude Bernard—celebrated today as the father of experimental medicine. Bernard routinely brushed aside the cruelty he inflicted on the animals he studied, arguing that the importance of their pain paled beside the nobility of his quest for knowledge. "A physiologist," he wrote, "is absorbed by the scientific idea which he pursues. He no longer hears the cries of animals, he no longer sees the blood that flows, he sees only his idea and perceives only organisms concealing problems which he intends to solve." Some of Bernard's colleagues, meanwhile, who were increasingly troubled by the cries of their own animal subjects, took to silencing them by cutting their vocal chords.

As for Bernard's wife, after suffering for years the awful sights and sounds of her husband's home experiments, she'd had enough. She took the kids and walked out. She then went on to establish a safe haven for lost and homeless dogs—the same ones her former husband was nabbing and torturing in his experiments.

Bernard's conviction holds the essence of a machine heart that on occasion still haunts us today: This belief that when it comes to exploiting "the other," and in truth that can mean either animals or entire groups of people, what a scientist or leader wants to find out or control justifies using whatever means are necessary to make that happen.

It wasn't until very recently that chimpanzees were no longer allowed to be imprisoned in eight-foot-square concrete cages. Among the most intelligent, sensitive, and social animals on Earth, these animals, like those rescued by Carole Noon, could be caged for thirty or forty years as the subjects of experiments ranging from massive infusions of psychotropic drugs and toxic chemical testing to being subjected to blunt-force trauma in studies of the damage from simulated automobile crashes. Finally in 2015 the US Fish and Wildlife Service listed all chimps in the United States as endangered, essentially ending biomedical and other studies on them. The National Institutes of Health followed with an announcement that all of its roughly three hundred chimpanzees were to be retired. Unfortunately, to this day many are still languishing in labs, waiting for sanctuary.

Not that there's been progress on every front. Though we may get squeamish at the thought of it, the worldview that allowed Descartes to go around kicking his dog, that encouraged a wholesale denial of respect to our fellow creatures and all but blinded us to the lessons of interdependence, is a cornerstone of how Western civilizations especially came to see the world. And we haven't quite grown out of it yet. Still, we're making progress. Contemporary Western science is beginning to confirm what indigenous people have known for thousands of years, uncovering an avalanche of new knowledge about our fellow animal travelers. Like the fact that it now appears that the neurons in the cortex of an elephant's brain—the part that allows higher cognitive abilities—may render it able to synthesize a greater amount of information from its environment than any other mammal on Earth, including us. Based on those findings, research neuroscientist Bob Jacobs reports that he and his colleagues believe elephants are "essentially contemplative animals." Unlike humans and chimpanzees, which are wired to make rapid decisions based on the stimuli around them, elephants seem especially well suited to curiosity and problem solving. Which dovetails well with what Ivory Coast tribes have said about elephants all along, praising the animals for their intelligence, cooperation, and moderation.

Then there's the clear scientific evidence that crows have an understanding of cause and effect rivaling that of first graders. And that many dogs can understand simple arithmetic. And that cockatoos can master picking locks in order to open boxes with food inside.

In recent years we've learned that a lot of animals possess what's sometimes referred to as self-consciousness—another way of saying that a creature is aware of itself. In 2012 a prominent group of neuroscientists signed on to the Cambridge Declaration on Consciousness, acknowledging the scientific evidence that many creatures, including wolves, are "aware" in many of the same ways humans are. They possess "consciousness, awareness, and intention"—qualities that from a human standpoint have been keystones of integrity and even morality for thousands of years.

We know, too, that some creatures even work together to build and maintain complex traditions among their families and kin groups—something we humans do routinely to shore up our sense of identity and belonging. Whales and dolphins display some two dozen recognized greetings and other bonding behaviors, the specifics fashioned by the preferences of a particular local group. Furthermore, those dolphins not only have unique whistles, which they use to identify themselves, but from what we can tell actually have names for one another.

With the bounty of new scientific understandings about our animal relatives, we have before us a golden opportunity, and at the same time a golden obligation, to heal those places where respect for other creatures has yet to be re-covered.

Let's get back to wolves. Part of the reason for our strong reactions to them—both positive and negative—probably has a lot to do with how uncannily similar wolf society is to our own. In fact, after studying their intensely complex social behaviors, some anthropologists and ethnobiologists suggest that when it comes to studying the evolution of human behavior, wolves may be a more appropriate choice for comparison than primates. But in one sense this isn't news at all. Little Red Riding Hood aside, thousands of tales from cultures around the world, from ancient Rome to the Lakota Sioux, have painted pictures of a primal kinship link between wolves and human society.

In the heart of modern Rome, at the middle of what is arguably one of the best classical art collections in the world, in the National Roman Museum, sits a stunning marble altar from around AD 200. Fashioned as a tribute shrine to Mars and Venus, it has carved into one side an especially compelling scene of a she-wolf nursing the twins Romulus and Remus.

Long revered as the mythical founders of Rome, the twins were said to have been born to a virgin named Rhea, who was mysteriously impregnated by the powerful god Mars. Now, all this took place at a time when there was an evil king in charge. Seeing the twins' astonishing beauty and strength, the king suspected right away that there was something of the divine in them. Afraid of the power they might one day wield, he quietly ordered that Romulus and Remus be drowned in a nearby river. When it came down to it, though, the henchman the king put in charge of the foul act couldn't bring himself to do it. Instead of drowning them he placed the twins gently into the river, where the water swept them away, soon to be caught on the roots of a fig tree. As luck then had it, a compassionate mother wolf found the snagged infants, lifted them from the water, and set about the chores of feeding and protecting them. Finally the gentle wolf and her charges were found by an equally gentle shepherd and his wife, who offered to take over raising the future Roman leaders to young adulthood.

Much happened for the twins as they ventured through childhood and into their public lives. There were sizable doses of intrigue and plenty of good adventures, including coming to realize their divine heritage. But not everything turned out well.

In a bitter quarrel between the brothers over where to locate the Roman Empire, Remus was killed, some say by Romulus himself. Romulus went on to establish the city of Rome, which is named for him, on Palatine Hill—an event Roman legend fixes to April 21 in the year 753 BC. He also established Roman law, and in his spare time served as a powerful general. With superhuman strength attributable to his inheritance from his dad, Romulus led the Roman army to victory after victory over a long list of enemies—even the Trojans.

All of that and more, say the Romans, made possible by that benevolent she-wolf that rescued the boys from a tangle of fig roots on the bank of a river. Divine offspring of Mars and the virgin Rhea, Romulus and Remus and the Roman Empire would never have made it were it not for a wolf. Which in the grand scheme of mythical poetry seems exactly right, since Mars himself considered wolves to be sacred.

Though the story of Romulus and Remus may be the most famous tale of infants being rescued and raised by wolves, it's actually just one of hundreds. In ancient Crete, Apollo had a child with Acallis, the daughter of a mean tyrant king named Minos. Fearing that Minos would be outraged by their coupling, they hid the baby in the woods. There wolves fed the infant and kept him safe until he, too, was rescued by herders.

In Ireland during the tenth century, the high king of Ireland was killed in a battle against his nephew. Afterward his wife, Achta, fearing for her own life, took their infant son Cormac and went into hiding in a nearby woods. One morning she awoke under a tree to find Cormac gone. Frantic, she and her servants scoured the woods to no avail. After Achta put out a generous reward for his return, a hunter finally found the baby boy in the caves of Keash. He'd been adopted by a mother wolf, who then cared for him along with her own pups. In fact, the hunter reported that when he found baby Cormac he was happy as could be, tussling about with his wolf brothers and sisters.

The list goes on. In Senegal the wolf was the first intelligent being created by the god Roog; in that culture the wolf was a great seer, one that it is said will remain on Earth long after humans have passed away. In Germanic, Celtic, and Mongolian mythic traditions, wolves were both companions of lost humans and messengers of the divine. And in equally ancient societies from southern Italy to Iran to Turkey to Chechnya, entire clans of people proudly referred to themselves as wolves, or at least as the descendants of creator wolves from the distant past.

Meanwhile the indigenous people of North America, from the Quileute to the Nez Perce to the Anisihinabe to the Zuni to the Pawnee (whom other tribes referred to as the people of the wolf), and plenty of others, not only saw themselves as descendants of wolves, but placed the wolves of this world in high regard as guides and teachers for how to live well.

As for me, after only months with the wolves of Yellowstone, I was fully in awe, amazed by what seemed a highly developed blend of reasoning and knowledge, and a kind of agency firmly grounded completely in cooperation. Those years I spent in the backcountry with the wolves of Yellowstone, I watched those animals in situations humans have faced across the centuries. How best to work together to secure food. How to keep squabbles in the family from turning into brawls. Engaging in the complex thinking needed to analyze danger, like suddenly meeting another wolf pack along a territorial border, and then reacting in a deeply coordinated way to make it out alive. In short, I saw them build and sustain flexible relationships based on trust and reciprocity.

Beyond the animals' strength and speed and power was a remarkably useful style of governance. The strong cohesion in wolf packs allows individual pack members to break off on their own now and then through the day. Young adults may leave the pack entirely for a time, roaming far and wide; if they don't end up finding a mate they may choose to rejoin their natal pack, and if so be thoroughly welcomed.

I observed in the Druid, Soda Butte, and Slough Creek packs of Yellowstone the extraordinary dedication wolves have when it comes to caring for their pups, with every adult sharing babysitting duties from May to September. These childcare duties were carried out in carefully chosen locations, often selected by the alpha female for how easy they were to defend from predators, their exposure to weather, and their proximity to good hunting. A portion of the pack's adults stayed with the pups, playing with them through every waking hour, while others in the group headed out on hunting forays, often bringing back meat in their stomachs, which they could then regurgitate for the pups. Then everyone switched off.

The old adage "it takes a village to raise a child" may well have arisen from careful observation of wolves.

Much of this behavior supports bonding in the pack—a key ingredient to survival. But bonding depends, as it does for humans, on personalities that can mix. Wolf packs don't do as well when they behave with one another like competitors. Plain and simple, cooperative personalities seem to create more thriving packs. More competitive wolves can have a tougher time getting food, since hunting requires extraordinary give-and-take. Lack of cooperation can also lead a pack to suffer higher predator-caused pup deaths, as well as more adult deaths from badly managed territorial disputes with neighboring packs.

For a very long time we've assumed that wolves were among the first animals adopted into human culture. An orphaned pup, the story goes, is taken into a camp and raised by the people who live there. But we might start also considering stories that describe things moving in rather the opposite direction. What if bold, highly affable wolves showed up around human encampments, relying on their wise discernment to actively seek us out as social partners? Not survival of the fittest, but rather, as Princeton evolutionary biologist Bridgett vonHoldt describes it, "survival of the friendliest."

Acknowledging our relations with animals can help create in us the peace and sense of belonging we long for. Spotting the twitch of a white-tailed deer during a walk in the woods, relishing the sight of birds on the feeders outside our windows, gazing into the eyes of our dog or scratching our cat behind the ears—all of it speaks of a sacred relationship.

Not surprisingly, the very same science that's helped us

better appreciate the intelligence and range of feeling in many wild mammals is now suggesting that the animals we raise for food have also evolved with impressive levels of consciousness. For example, despite having been domesticated from wild boars about nine thousand years ago, pigs show a strong innate ability to perceive and engage the world around them. They have excellent long-term memory, and they can prioritize the most important of those memories. They can learn both verbal and gesture-based instructions. And pigs love to play—be it pushing beach balls with their snouts or exploring new environments or jumping into the water for a swim just for the sheer fun of it—in short, showing the kind of playfulness long associated with the most adaptable, intellectually complex animals. Likewise cows, as most any rancher can tell you, are extremely sociable, have distinct personalities, react to stresses in ways not unlike humans, and can remember both people and other creatures who treat them either kindly or harshly.

Almost none of us would advocate going back to treating animals in the harsh, unfeeling ways they were sometimes treated in the past. But at the same time most of us have been remarkably willing to avert our eyes and hearts from the well-being of the animals that die or suffer to give us our daily bacon and eggs and glasses of milk.

Yet as early as the seventeenth century, brilliant physiologist Robert Boyle declared cruelty to domestic animals a blasphemy against creation. As it was for the members of the Massachusetts Bay Colony in New England, who in

1641 outlawed "Tirranny or Crueltie" to domestic animals. Even in medieval Europe, when not a lot of thought was being given to animal welfare, scholars and theologians alike strongly advised against leveling cruelty toward other creatures because of the terrible effects it had on human character. And as Charles Darwin let us know 150 years ago, "The lower animals, like man, manifestly feel pleasure and pain, happiness and misery."

This is one of the most uncomfortable animal conversations in all the world. The domestic creatures we raise for food have considerable intelligence. These beings, like us, like redwoods and bonobos and jaguars and eagles, are not simply widgets. How do we reconcile that fact with the dramatic rise over the past half century of raising them in harshly artificial conditions—in so-called concentrated animal feeding operations. Does affordable food really require subjecting thinking, feeling creatures to astonishing levels of boredom, confinement, and even pain? Would taking steps to improve their lives still allow food to be affordable enough for families to put dinner on the table?

Only about sixty years ago the majority of farms in the United States were diverse enterprises. Farmers raised a variety of crops and animals, the former feeding the latter, with most of the pigs, cows, sheep, and chickens enjoying at least a modest dose of life under open sky with their feet on ground. But the Second World War brought developments that would soon change forever the world of farming. The same nitrate–synthetic ammonia fertilizer pioneered in

Germany in the early twentieth century to make crops grow (and later, to make bombs) was by the end of the Second World War being used routinely in the United States too. That fertilizer, along with hybrid seeds, pesticides, and ever more powerful planting and harvesting machines, increased productivity. By 1970 US corn farmers had increased their yields by 500 percent. In other words, all of a sudden there was lots of corn.

More corn, in fact—and more of other grains too—than the markets could handle. And that made grain cheap, suitable for feeding to animals. This set the stage for explosive growth in animal confinement agriculture. Today in the United States just four large corporations produce 81 percent of the cows we eat and raise for milk, 60 percent of the hogs, 73 percent of the sheep, and half the chickens. Hundreds of thousands of cattle are raised in crowded feedlots, fattened on grain, while the calves used for veal are often chained as newborns in tiny stalls where in some cases they can neither turn around nor lie flat. Hog mothers are sometimes confined ten months of the year, or, if artificially inseminated, can be continuously confined throughout their lives, much of the time in pens too small to turn around. Meanwhile egg-laying chickens often spend their lives in wire "battery cages" measuring about twelve inches by eighteen inches, sometimes sharing that tiny space with several other birds.

On one hand this shift to mass confinement has had big consequences for the environment. The use of antibiotics, popularized in the 1950s when it was discovered that such drugs made cattle put on weight, became more widespread as animals were packed together in feedlots, where they're more vulnerable to infections and disease. It's in part because there are so many antibiotics in our food and water from animal agriculture that attacks on humans by antibiotic-resistant bacteria are rising at a frightening rate. In the United States alone, fighting such infections now costs more than two billion dollars a year.

Meanwhile runoff from concentrated animal production facilities is a big reason why 173,000 miles of waterways in the United States are classified as "dead zones," too polluted to sustain aquatic life. More than half the soil erosion in the country is caused by animal farming, as well as a third of the nitrogen and phosphorus pollution in our drinking water. To put that pollution in sordid perspective, consider that waste slurries from hog production facilities are about seventy-five times stronger than raw human sewage.

Then, of course, there's the matter of the animals themselves. The struggle to bring more humane conditions to industrial animal agriculture has now been going on for well over fifty years, first sparked in 1964 when Ruth Harrison of the United Kingdom published the book *Animal Machines*. What Rachel Carson's *Silent Spring* did for curbing the use of pesticides on food crops in the United States, this book did for farm animals in the United Kingdom. Harrison pointed to two root causes of the animal suffering in domestic livestock operations—changes that allowed things to go terribly wrong. First, she said, the relationship between individual farmers and their animals had unraveled, what with the animals no longer a daily part of life on the farm. And second was the shift to housing the animals in large, windowless warehouses—totally concealed from public view.

Within a year of *Animal Machines* hitting the streets, a group of leading biologists, veterinarians, and animal scientists got together to help the government launch a response. In what for the time was an enormous step, the United Kingdom declared that the health and well-being of farm animals must take into consideration not only an animal's physical condition, but also how it's doing mentally. In time the British government endorsed something called "the five freedoms"—essential liberties guaranteed to all animals.

*Freedom from hunger and thirst.*
*Freedom from discomfort,* to be achieved in part by providing adequate shelter and resting places.
*Freedom from pain, injury, or disease.*
*Freedom to express normal behavior*—including sufficient space for an animal throughout her life, as well as the company of other animals of her kind.
And finally, *freedom from fear and distress.*

Before long, the extension of those freedoms was showing up in other countries around the world.

Notably, the enforcement of such guidelines resulted in relatively little increase in costs to consumers. Still, not everyone came on board. The United States was resistant, mostly toward the call for farm animals to be given a semblance of "normal behavior." Many advocates of large-scale animal confinement, adding to their economic arguments, claim that such practices actually promote animal welfare by minimizing disease and by treating illness quickly.

This, then, is the tension: between a fully relational ethic across species and an ethic forged from a fiscal advantage to humans.

Back in 1997 a group of top scientists for the European Union looked at the effects of using tiny gestation stalls in pig farming, where sows are often chained at the neck in pens too small for them to turn around. The group concluded that even in the best circumstances the pigs were seriously compromised, showing all the signs of severe depression. As a result, the practice was phased out. Soon afterward a similar review was made in Australia, but this time the researchers—driven by the physical-health / commodity-enhancement argument—came to the opposite conclusion. To this day the United States generally ignores the European study and opts instead for the Australian version. And widespread use of gestation stalls continues.

The more we learn about the minds and emotions of animals, the clearer it becomes that being free of disease and predators is only a small part of the story. Seventy years ago, American comparative psychology researcher Harry

Harlow set out to create an entire group of disease-free rhesus monkeys. In part his plan had to do with the idea that when it came to measuring behaviors in the monkeys, it would be helpful to make sure that no illnesses were altering the results. Seemed like a good idea. To accomplish his goal he took babies from their mothers just after birth and put them in separate cages. While they could hear and see other monkeys in other cages, the young were denied any contact with one another. And Harlow was right. In measurements of illness and disease the monkeys did well. Yet to his horror, he soon came to see that the animals were severely emotionally disturbed.

"The laboratory-born monkeys sit in their cages and stare fixedly into space," he noted, "circle their cages in a repetitive stereotyped manner and clasp their heads in their hands or arms and rock for long periods of time." Exactly the sort of behavior humans engage in following severe trauma. Harlow's monkeys seemed fine in terms of physical health. And yet they were far from joy or comfort, bereft of contentment.

"I think using animals for food is an ethical thing to do," says noted animal scientist Temple Grandin. "But we've got to do it right. We've got to give those animals a decent life and we've got to give them a painless death. We owe the animal respect."

Some people are choosing to eat meat a little less frequently, others not at all. Some make it a point to buy meat and poultry certified as being raised in more humane fashion. All well and good. But we also need to participate in bringing simple, affordable changes to industrial agriculture—ensuring that it better respects the millions of animals that each year give their lives to feeding the people of the world. And even as things get better, we need to understand there will always be a powerful temptation to make decisions based solely on the bottom line.

Incredibly, in 2017 while debating the Brexit Bill, the Tory government of Britain reached back across some four hundred years and declared again that animals have no emotions or feelings—countering the work of Ruth Harrison and the hundreds of biologists and veterinarians who came in her wake. They then went on to vote against transferring into the UK law the EU legislation recognizing that animals have sentience and can feel pain.

Recommitting ourselves to greater respect and compassion for farm animals—and compassion really is deeply woven into our nature—allows a healing of the age-old division between us and the rest of creation that caused us to turn a blind eye to suffering of all kinds. Being good to any creature, at any time, will help us find our hearts again.

It's October in the Northern Rockies, but warm weather as of late speaks more of summer than of fall. Still, for those out walking greater Yellowstone, one has only to cock an ear to the hills to know that the seasonal changes are under way. Ringing down these hills are the piercing, reedy bugles of bull elk, deep in the season of the rut. By the end of that time the bulls will have expended a staggering amount of energy in their efforts to mate, sometimes eating little or nothing along the way.

What the big bulls have going for them this autumn, though, is the hearty, snow-filled winter that came the year before—a condition that's becoming less common in these years of climate change. Like other ungulates in late spring, elk push up the slopes of the mountains into the high country with their noses held nearly against the receding snow line, nipping at the youngest, most nutritious graze of the year. In years of good snowpack an elk can ride this "green wave," as biologists call it, all the way into August. In low-snow years, on the other hand, that wave crests early, often in the middle of July. Such an early crest leaves bull elk going into rut with considerably less strength than would otherwise be available to them.

Wolves are by nature always paying close attention. They've figured out that when a rut happens in years of little winter and spring snow, bull elk will by the end of rut be more thoroughly exhausted. In other words, the powerful animals that would normally for wolves be extremely dangerous prey, too risky to even try taking, may be vulnerable. But probably not this year. This year, because the bull elk has been well fed with vegetation watered by snowmelt, his essential strength will stay intact. As robust as winter itself.

You can't spend much time in this wild heart of Yellowstone—the so-called American Serengeti—without noticing the unique genius of the animals that make their homes here. Animals driven not merely by how brilliantly evolution has tuned them to their environment, but by the conscious choices they make in any given moment. In the wolf world, both predator and prey are constantly assessing risk. For the elk, that means scanning a nearby wolf pack to see if it's just passing through or if any of the pack members are in the specific physical postures that signify hunting mode. The wolves, for their part, if in fact they're hunting, will first prod the elk herd into running. By watching the herd move they'll be able to figure out if any of the elk look vulnerable. Maybe breathing hard—sometimes a sign of pneumonia. Or moving with a limp. If none of the elk slow or stagger, the wolves often will move on, thus sparing themselves the increased danger of being kicked—of having bones broken or even being killed—that can happen when trying to bring down a healthy animal.

And all the while the raven glides overhead, watching the drama from above. She and the wolf packs have their own kind of synchrony—the raven actually showing the wolves the way to the elk herds, well aware that if the wolves make a successful kill it will allow her to feed too. The raven, you see, can't tear through the carcass of an elk with her beak, so the wolves do it for her. The wolves are fed, as is the raven. For that matter the coyotes in the timbered foothills have their eyes on the ravens too, probably thinking up some clever plan to try to feed on any leftovers without being trounced by the wolves.

When an individual elk does fall to the wolves—something that happens only in about one out of every five hunting attempts—the elk herd may well be left stronger. Indeed in the years to come wolves are likely to show themselves as a powerful control agent in the face of rising incidents of chronic wasting disease. This neurological disease in elk and deer, rapidly passed from one animal to another, eventually kills by attacking the brain and nervous system. As it happens, one of the early symptoms of chronic wasting disease is profound lethargy. And that lethargy means infected animals are likely to be the first ones chosen by wolves as safe prey.

In seven months there will be a flush of new life here. Bison calves jumping around like they have pogo sticks for legs, playing always within a ring of adults who circle them to keep them safe. And elk calves, too, trotting with their mothers and the rest of their herds only a few days after birth. Bears will be out of their dens, ambling about sniffing the air for the scent of carcasses from elk and deer who perished the previous winter. And also being lazy, which bears are very good at, stretching and purring in the deep shade of their day beds, waiting for the heat of the afternoon to pass.

All of them, like us, trying their best to make a living,

moving through the wild and often uncertain circumstances that call them forward into their lives.

Weaving the dignity of animals into how we act in the world means expanding our own devotions beyond family and friends, beyond neighborhood and country, beyond humans even, to create a new kind of loyalty to all living things. A loyalty that—in a true honoring of nature's lesson of interdependence—is less about saving any one species than it is about saving us all.

And as extraordinary as it may seem, wolves—and elephants and dolphins and whales and ravens, and yes, even cows and pigs—can help show us the way.

# We Live on a Planet with Energy Beyond Measure, Yet Life Doesn't Waste a Drop

Every aspect of our lives is, in a sense, a vote for the kind of world we want to live in.

—Frances Moore Lappé

Some years back, science writer Oliver Morton came up with a fascinating mind-bender to help us better imagine the staggering amount of energy contained in the sunlight that falls on Earth. He compares this sun energy to a river—and more specifically, a waterfall. First Morton asks us to hold an image in our minds of Niagara Falls. And next, to then grow the height of those falls by a factor of twenty. That would grow the actual 187 vertical feet of Horseshoe Falls to about 3,700 feet, which is about three times as tall as the Empire State Building. Next Morton tells us to imagine the flow at the falls increasing tenfold, which as best as I can figure would bring it to more than twice the volume discharged each second into the Gulf of Mexico from the Mississippi River.

But we're still not even close to the energy given us by the sun. So Morton has us widen the falls. A lot. Until the cascade stretches around the entire equator. What we end up with is a staggeringly massive waterfall three times as high as the Empire State Building, encircling the planet at its midpoint, every second pouring billions and billions of tons of water over the lip.

That's an astonishing amount of energy. To put it another way, more energy falls from the sun in just an hour and a half than all the energy humans consume from all sources in an entire year.

Given that all life on the planet comes directly or indirectly from this sunlight, it might seem odd to learn that nature puts an enormous emphasis on efficiency. Why would there be efficiency in the face of such abundance? Yet every leaf on every tree is tuned down to the subatomic level for gathering as much light as possible into every pore. Every creature, meanwhile, is flush with an ease of movement and spooling of body functions, showing an efficiency that's long been the envy of physicists, architects, and design engineers. Indeed, much of today's cutting-edge technology is an attempt to mimic biological efficiency, whether we're talking about bullet trains or vaccines or water filters or wind farms.

As it happens, nature loves efficiency because living beings can only capture so much of the energy available to them. Also, whenever a life-form takes in fuel—whether directly from the sun, like a blade of grass, or indirectly, like when the impala eats the grass, or the lion eats the impala— a lot of energy is required to turn it into something usable. So yes, we live on a planet with an endless supply of energy. But from a biological perspective each living being has been fated to work out a puzzle where the name of the game is to make the best use of the essential nourishment sunlight offers, without wasting a drop. In the most basic sense the purpose of life is to keep more life flowing—unfolding as many trees, as many butterflies, as many wild roses as the ecosystem will allow. That takes efficiency, and those species that prove best at it live the longest, and live best.

Jack Gladstone, a Grammy-nominated Blackfeet musician living at the edge of Glacier National Park, has an interesting way of framing all this. He says that what he focuses on in music is really what drives all of life. Every species. Every ecosystem. "It comes down to three things," he says: "harmony, balance, and rhythm." Whoever and whatever masters those will have a successful life.

He's onto something. All of nature does in fact turn on those three things, and each one can be understood in terms of energy. Harmony, for example, can be seen as not pushing against whatever's happening in any given moment, but rather adjusting to move with the flow. A migrating duck caught in a fierce headwind, instead of exhausting itself, may land and wait for more favorable conditions. The second quality, balance, can be understood as striving for an equilibrium between energy coming in and that going out. That same duck will stop and fill up on grain well before her body has to start breaking down muscle cells in order to feed itself for the exertion of flight. And finally there's rhythm, which includes the daily, seasonal, and even lifetime cycles that steer us all—creating alternating beats of strong activity and calm restoration. A bear, facing cold, snowy weather and therefore the end of easily available food, will crawl into a den, lower her metabolism to a tiny fraction of what it was, and sleep until the first signs of spring.

No matter where we look, we find nature using harmony, balance, and rhythm—the artistic infrastructure of conservation. Take, for example, the humble, sweet-faced sloth. Not surprisingly for a creature that lives in trees, the sloth is all about the leaves. Now, on first glance leaves might seem like the perfect daily blue-plate meal. After all, in a lot of the world, leaves are found in staggering abundance, just hanging there for the taking. But there's a catch. Actually a couple of them. First of all, leaves aren't all that easy to digest. What's more, they tend to not have all that much nutritional value. The truth is it's darned hard to squeeze out the calories a mammal needs for moving through the world just by eating leaves. Even at the poky pace of a sloth.

The optimal leaf-to-movement ratio is precisely the puzzle both the two- and three-toed sloths of Central and

South America set out to master many millions of years ago. The two breeds of sloth—varying by, yes, their toes— actually set out separately on their investigations, coming from completely independent family lines and living in communities that had no contact.

Each lived around lots of food, leaves all over the place, but again the scant amount of nutrition posed a problem. What to do? One thing that worked out really well for both species was to fashion a set of feet perfectly made for hanging upside down on tree branches, surrounded by food. And within that one adaptation is a big evolutionary nod to energy conservation. Hanging upside down significantly reduces calorie consumption compared to what would be needed if you were spending your day balancing on top of or between tree branches. Both types of sloth also perfected small shoulder blades and long arms, a design that allows for hanging around in one place for extended periods while remaining within easy reach of the next bite.

Sloths lowered their body temperature too, drastically dropping calorie requirements. While most mammals, including humans, keep their insides at around a hundred degrees, varying little unless they're sick, sloths adjust according to demands, able to shift their body temperatures across a range from seventy-four to ninety-two degrees. If you think the lower end of that range might make a critter cold, you're right. Which is why sloths not only live in tropical jungles, but spend a fair amount of time warming themselves in full sun at the tops of the trees. When it's nap time they curl up in tight balls in the forks of branches to recirculate their own body heat—yet another move that helps conserve calories.

And sloths have figured out even more astounding ways to conserve energy. Like the fact that they sublet their fur. The short-hair coats of these animals are home to a variety of small organisms, from beetles to moths to blue-green algae. The algae specifically thrive on the moisture in sloths' hair to such a degree that during the rainy season the sloths can turn a rather lovely shade of green. On one hand this is a good way to hide in the trees from would-be predators. But more to our point, every time a sloth licks her fur she takes in a bit of that algae, which provides a little extra fuel.

Finally, there's that famous slow motion—and it's true, sloths really do nothing quickly—another way to save loads of energy. Even digestion in a sloth proceeds only in the most sluggish fashion. A big meal can take a full month to work its way through the animal; the resulting bowel movement—and those happen only about once a week— can reduce its weight by more than 30 percent.

The long and short of it is that sloths may look awfully lazy. But in reality they're energy-efficiency magicians, having come up with a unique weave of balance, rhythm, and harmony. One that's kept them going long after their distant cousins, the giant ground sloths, disappeared from the face of the Earth.

Right now on Earth we think there are about sixty thousand mammals, birds, reptiles, and fish, and more than three hundred thousand different kinds of plants. All of them have survived by building lives of exquisite efficiency. Birds fly, lions run, bees make honey, flowers unfold, apples ripen, fish swim, trees reach for the sky—all with expenditures of energy trimmed to the subatomic level. Your body does this too. Each of your cells has come up with a neat way of burning sugar from the food you eat—doing so in tiny steps. With every step a little energy is released, some of which gets stored in special carrier molecules, available for later use. Technically, if all that sugar got burned at once, it would produce the same amount of energy. But energy coming in a big burst like that would be more than your cells could actually make use of. Which means a lot of it would go to waste; lost—as all energy finally is—in the form of heat. It's sort of like a person trying to keep herself warm through an entire night at the edge of a campfire. She'll fare better if she feeds the logs as they're needed, rather than dumping the whole woodpile into the flames at once.

So what about the other end of the spectrum from a sloth—a creature like, say, the hummingbird? A being that darts about with its wings whirring and blurring at a rate of roughly fifty beats a second? Clearly a part of this bird's strategy is to eat a lot—really, a lot—taking in two or even three times her body weight every single day. In fact, if you projected the size of a hummingbird's body to your size, as a human you'd be consuming that superfood called nectar at a whopping 150,000 calories a day—about seventy-five times the number you need to keep your human form going.

Beyond the hummingbird's strategy of eating a lot, there's no end of efficiency in play. For one thing, these exquisite little birds have shaved their weight to the absolute minimum. To lighten the loads they have to cart around from flower to flower, they've done away with the usual downy feathers that most birds use to keep warm. Instead they're able to enter at will a state of extreme torpor, mostly at night, lowering their body temperatures by nearly fifty degrees and dropping their heart rates from a stunning five hundred beats per minute when flying to an equally stunning fifty. At these times of rest, their breathing comes nearly to a standstill.

Then there are those creatures that play the energy game by modifying their homes. Bees, for example, have embraced complex geometry in building their hives. In fact, they've mastered the most efficient storage scheme known to humans. Tiny wax balls exuded by workers are cooperatively fashioned into the slightly bulging hexagons of a honeycomb. Even the ancient Greeks considered the design brilliant, though it would take two thousand years before University of Michigan mathematician Thomas Hales, in a fairly staggering 250-page mathematical proof, showed that the bee's hexagon designs set the absolute lower limit of this architectural geometry—using the smallest amount of surface area for the maximum amount of storage. Any other geometric design—say, triangles or squares—would require both more space for the same amount of storage and more energy for producing the wax building material.

In these pages we've been investigating what makes nature actually work—how it thrives, what makes it resilient, and what those ingenious natural designs and processes might suggest about the choices we make for bettering our own lives. To that end, nature's emphasis on efficiency can be a great teacher.

A hundred years ago the brilliant Hungarian biologist Ervin Bauer listed three fundamental requirements for life. One of them was that every living thing has the ability to use the free energy it holds to move or think or increase its growth—basically, as he put it, to "do work." Of course, that isn't true in the nonliving world, where from an energy standpoint things "behave" in a more straightforward manner. They abide by those basic rules of physics you learned in school, first laid down by Isaac Newton—like the fact that an object will remain at rest or in a uniform state of motion unless it encounters an external force. And further, that for every action there's an equal and opposite reaction. A baseball comes off a powerful hitter's bat, and depending on the speed of the bat and the ball, and the angle of each, it rockets toward the back of the stadium wall in a very predictable path, coming to rest, its energy spent, maybe in the mitt of some eager fan. End of the story.

But living things are different. We're continually investing energy to create more of it. Life, when you think about it, is really about creating more life. And yet again, there's no end of efficiency when it comes to carrying out that task. Maybe there's something we can apply from that in how we think, the decisions we make, how we go through our lives.

What if we frame the idea of thought influencing our well-being around the question of whether or not those thoughts are a sound use of energy? We already know that thinking causes reactions on a molecular level. After all, you *will* your body to do hundreds of things every day: You decide to get out of bed in the morning, and will your body to do it. You raise your arm to wave to a friend, or bend down to scratch your dog behind the ears. But even more idle thoughts—ones with no physical aim—likely put into motion cellular-level responses that can reflect either calm or agitation. A thought can lead either to a quiet, efficient expression of energy or to something more chaotic, more wasteful.

It's been shown time and again that positive, joyful ruminations lead to better health. And good health—a body and mind in balance—does indeed suggest a kind of efficiency. At the same time, persistent anger and frustration, lingering thoughts that are contrary to our notions of who we are and what we need, can compromise us. Modern neuroscience suggests that we play a very big part in creating reality—that much of what we think is going on in the outer world is in fact really about what's going on in us. That same neuroscience tells us that thought habits can be changed, and changed at almost any time in our lives, whether we're young or old. What if we consider nature's love of efficiency, this default it has for minimizing effort, as a cue for trying to live with less mental struggle, to create more vibrant lives by using our mental energy wisely?

Over the course of a lifetime, and often beginning early in childhood, we can create anxieties that burn lots of energy without getting us any closer to what we seek. Too often we block ourselves from being able to rest in our deepest natures. We worry, Am I thin enough? Good-looking enough? Smart enough? Do people think I'm successful? Will people approve of whom I love? Everybody feels such concerns. But by lingering too long in rumination, by giving it too much energy, we end up never quite able to disentangle. We rob ourselves of our own agency, creating levels of unease that in the end may deny us, our friends, our family, and even the world at large the real gifts we have to share.

When Albert Einstein kept intentionally disrupting his usual logic patterns in the woods of Princeton, when Walt Whitman went to the woods of Camden and opened himself to whatever physical and mental healing might happen there, they may well have been prompting biochemical responses that allowed them to get at least some of the very thing they sought. As if they were engaged in a kind of "inner feng shui," a mental version of the ancient Chinese practice that seeks to align physical spaces with the energies said to link humanity to the wider natural world. One important way we humans can emulate nature is to ease off our compulsive thinking, to put down our tendency to admire the problem and second-guess every solution. Rather than succumbing to inner frenzies that interfere with what we really seek, we can become more effortless, emulating those parts of the natural world rich with calm and quiet.

This may be one of the greatest benefits of nature—that it can lift us above the inner fray, letting us better access deep feelings of being aligned with the bigger world beyond our egos. University of Illinois biologist Dr. Frances Ming Kuo recently called the range of specific health benefits tied to nature "startling," highlighting a list that includes "depression and anxiety disorder, diabetes mellitus, attention deficit hyperactivity disorder, various infectious diseases, cancer, healing from surgery, obesity, birth outcomes, cardiovascular disease, musculoskeletal complaints, migraines, respiratory disease, and others."

Nature has also been shown to be a fundamental ingredient of so-called attention restoration. Through something psychologists call "soft fascination," our time in the natural world has the effect of easing the fatigue that comes from the long periods of focused attention common in our daily lives, allowing us what amounts to a mental and emotional reboot, one that allows us to go back to tasks that require directed attention free of our prior fatigue.

Nature's effect on rebooting attention has also been supported by empirical research on attention deficit disorder. Years back, as I was writing about a group of struggling teens in a compassionate wilderness therapy program, I was astonished to see the effects of the wild on kids with severe ADHD. Young people who on arrival could barely focus well enough to hold a two-minute conversation were within two weeks able to gain focus enough to complete complex tasks like building bow-drill fires—often for hours at a time—and more significant still, to then teach that skill to others, which often required extraordinary patience. A number of them, under the guidance of the supervising therapist, went off Ritalin entirely, this despite having been on the drug for most of the previous decade. Some stayed off the drug even after they returned home. One such girl suggested it was because she could still access strong memories of the deep calm she felt in nature.

At Stanford University, meanwhile, researchers have documented actual changes in brain activity from people walking in nature versus walking that same amount of time on busy city streets. Specifically they were looking at neural activity in the part of the brain that tends to ruminate, focusing on worries and negative emotions—a region known as the subgenual prefrontal cortex. Out in nature, that rumination decreased significantly. The natural world helps us regulate wayward emotions, which in turn lessens our anxiety. While these benefits can last from days to weeks, it's worth noting that they seem to accrue. In other words, the more we turn to nature across our lifetimes, the longer and deeper are the positive effects.

The beauty and mystery that embrace us in the natural world, that web of connections going on all around us, somehow nudges us toward a greater presence of mind— this sense that nothing real is ever happening other than what's unfolding right now. It's an idea fundamental to many meditation practices—that in any given moment nothing could be other than it is; and as it happens, that notion is a lot easier to hold on to, to breathe into, out in the grass and the sun and the trees.

Mental efficiency for us humans, then, might look something like this: Accept what's going on because that's what's going on. Act in whatever way the circumstance calls for, if in fact it calls for anything at all. If you do take action, don't waste mental and emotional energy either second-guessing yourself or getting overly attached to the result. And finally, try to take to heart an idea that's been around for well over two thousand years, which suggests that taking care of the self on the deepest levels means being grateful for the fact that you already have everything you need. From what we can tell, every one of these essential perspectives may root in you, and grow more quickly, when fed and watered by the natural world.

The next time you have the chance, maybe sometime in mid-November, turn your gaze to the sky to see if you can catch the ancient spectacle of wild geese winging southward—flying through skies just ahead of the cold slap of winter. You'll probably hear them first, then see them, cruising at forty, fifty, even sixty miles per hour. They are maybe only a few hundred feet up—or maybe thousands of feet above you, tiny specs drifting over a patchwork of brown grass and stubbled fields and the freshly bared arms of maples and ash and oaks.

Were you somehow able to ascend into the skies and be among those geese, one of the things you'd notice right away is that the lead bird is in a delicate dance of body adjustment, always minimizing the energy she's putting out by leaning into and out of even the most subtle wind currents. And if your day of flying happened to be in the uplands— maybe the White Mountains of New Hampshire, or the Adirondacks, or the Rockies or Sierras—you'd also see a kind of exquisite roller coastering in the flock as it maneuvers across the terrain. Whenever possible the birds drop into the lower valleys, taking advantage of greater air density at those lower elevations, which allows them to catch more air with every wing stroke. That means they use less energy. Then when the time comes, they rise together to cross the high divides. And then there's the behavior you've probably noticed dozens of times from the ground—something called "drafting," which plays out in that familiar V formation used by birds around the world. Drafting allows followers to benefit from "aerodynamic washup," not only avoiding headwinds but actually gaining lift from the birds in front. On one hand it's about flying in just the right place. But it's also about stroking your wings at just the right time. If you were up there right next to the birds, you'd see that it's a beautifully refined technique, perfectly tuned. So much so that researchers estimate drafting allows geese to fly about 70 percent farther than they could if they were traveling on their own. Each goose will take the lead slot for about the same amount of time, and then will spend the rest of its journey drafting behind other birds—rotating from the head to the back and then moving up again, bird by bird by bird.

This avian dance is accomplished in part through a steady stream of communication—that throaty honking you hear—which includes routine check-ins between the leaders and those in the back of the flock to keep track of how everyone's doing. Should one of the geese start struggling—or, say, during hunting season become injured—typically two other birds will fall out of formation with it, follow it to the ground, and remain there until their companion either recovers or dies.

So far, we humans have taken some of what we've learned from these wild geese and applied it quite literally. That's how we ended up with World War I military pilots mimicking flying-goose patterns in their formations, putting aerodynamic washup to use for much the same reasons as the birds do: to gain lift, to reduce headwinds, to minimize the energy required to travel, and to keep each plane in visual contact with the others in case of trouble.

So, too, have we learned similar lessons from, of all things, schooling fish. And in particular, from the beautiful, enchanting flutter of large groups of fish moving en masse through the water, the trailing fish effortlessly keeping perfect time with the leaders. This underwater dance is more than beautiful art. A single fish propels itself by twisting its body to create small eddies, or whirlpools. The trailing fish wrap their bodies around those disturbances of water, and by doing so get pulled along. Learning from this particular slice of nature, a group of students from the California Institute of Technology decided to assemble vertical wind turbines in tightly packed groups, like schools of fish. That bit of biomimicry, as biologist Janine Benyus points out, then produced a phenomenal tenfold increase in wind power.

It's a powerful thought to imagine what it might look like to live in a world where families, workplaces, cities, and nations were more fully aligned with the idea of living efficiently, taking only what we need. To participate deeply in something the great Chinese philosopher Lao-tzu observed about the natural world, which is that "the way of nature is to take from what has excess in order to make good what is deficient."

With these very ideas in mind, cultures around the world have engaged, likely since prehistoric times, in what many indigenous people today call a "giveaway." If you go to a traditional Native American wedding or a naming ceremony or funeral, it's likely that a giveaway will be a part of the event. In many tribes a blanket is spread with unwrapped gifts of every sort, and each guest is invited to come and take one thing. Often elders go first, then veterans, women, little children, older children, and finally the remaining men.

The fundamental energy behind the giveaway, openly expressed, is gratitude for the fact that life has given us all we need: water, air, food, shelter, warmth, community, and beauty. That gratitude, in turn, is a powerful midwife to the ethic that what we accumulate beyond the basics should be shared with those in need. By individuals giving to the needy, the system remains both diverse and "energetically balanced," and therefore more likely to thrive. Today's needy, after all, may well be tomorrow's givers.

Curiously, Dr. Mary M. Clare, a social scientist who's spent decades observing and supporting community health, has found that the groups most likely to consistently engage in this "natural sharing" are those that have the least: homeless people. Farmworkers living in migrant camps. "It's in those communities," she notes, "where we often find generosity at its greatest." By all appearances, humans have a natural inclination to serve the interests of the whole by nourishing the well-being of each individual. This kind of generosity can be understood as deep efficiency. It benefits the entire community by keeping those who have the most connected to those with the least, thereby bonding and profoundly strengthening the community.

We touch the power of this sharing system when we who have plenty feed the hungry or help house the homeless, when we give either friends or strangers a hand up when times are hard for them. When we offer our time and talents as gifts with no strings attached. Every day brings a chance to back off a bit from equating a person's worth to whether they're able to pull themselves up by their bootstraps, and to focus instead for a while on helping them get the boots. We can be grateful. We can use what we need and pass along the rest.

As it turns out, this gratitude—which, again, is the primary energy of the giveaway—can have powerful palliative effects. A recent study at UCLA found that the simple act of *looking* for things to be grateful for causes a significant boost of dopamine in the brain stem—essentially mimicking the effect of antidepressants like Wellbutrin. At the same time, gratitude has been found to cause an increase in serotonin production in the anterior cingulate cortex of the brain. Precisely the effect of Prozac. This sharing isn't just a one-way thing. It's circular, benefiting both the grateful recipient and the grateful giver.

What a blessing, says an old Islamic tale from Morocco, that there have always been giving people, generous men and women with hearts as wide as the sky and brilliant as the sun. One among these was Princess Hatim, daughter of a sultan. Now, Princess Hatim was wealthy beyond most people's grandest dreams, yet not a single dirham did she keep, preferring instead to give her riches to the poor. Though some of the members of her own family thought her foolish, they didn't interfere. It was her wealth, after all, and she was free to do with it what she wished. But great trouble arrived on the day her father found out that in addition to giving away her own fortune, she'd also been giving away his.

Hatim hadn't asked her father, nor had she ever mentioned that she was taking his gold and giving it to those in need. Nor, in truth, did she show any remorse when her father confronted her about it. "What would you have me do?" she asked, mystified that he would be upset. "Would you have me close my hand in the face of sickness and misery?"

But her father saw the matter very differently. In the end he had to conclude that because his beloved Hatim had stolen from the royal family, she must be punished like any other.

"You may choose death," he told her sadly. "Or you may choose exile."

After much anguish, Hatim told her father she would

take death. How could she, after all, a daughter of the Maghrib, survive in a place far away from her beloved homeland?

And so it was. But Allah was looking down from above on that fateful day. Granting favor to Hatim for having been so generous, He decided to turn her into an almond tree, the finest of all the trees. How appropriate. To this day the almond tree continues to give gifts to the people: nuts and oil to feed those who are hungry. And in spring, flowers so beautiful they cannot fail to bring joy, lifting even the most troubled heart.

Honoring the basic realities of existence on Earth, and in particular the connections that sustain us, has the effect of strengthening our commitment to protect those relationships. Often these days, struggling to stay calm and centered above the noise, holding on against what feels like a raging river of distractions and demands, we can hear the call to serve the energetic needs of the life around us as another burden. This is because we diminish the call by seeing it only as energy going out. But these fundamental affirmations of life, these essential energy exchanges, hold the very vigor we seek. Nature doesn't lose energy by virtue of relationship. It gains it. The efficiency of nature is always in service to life creating life.

While many native cultures stay connected to this truth with customs like the giveaway, the ancient Greeks made it a prominent point of their stories. They were fond of telling tales about a group of immortals known as the Titans—a collection of badly behaved deities who came to be known for their insatiable appetites for unfettered power. No doubt the Titans would've scorned an idea like the giveaway. In fact, had there been such a thing as bumper stickers back then, the one on the backs of the chariots of the Titans would've read, NO LAW. NO LIMITS. The Titans were excessive. Hugely chaotic. And they were considered both highly seductive and extremely dangerous to flowering young humans—youth who were overflowing with all the eager life-force of the bud, but with none of the patience or knowledge for tending its unfolding.

To counter the seduction of the Titans, the Greeks wove other stories that celebrated deities whose job it was to guide youth toward more generous, more thoughtful ways of living.

Chiron, for example, was the wise guide for an extraordinary array of heroes from Hercules to Achilles. Chiron's wisdom and teachings arose from his sense of the *balance* of nature—specifically, redirecting the impulsive, titanic energy of immature humans into service of the larger community. Chiron showed his students how to use the genius of nature's interdependence as a guiding principle for their own lives. Through him they learned to marshal the raw power given to them at birth and apply it to shoring up the resilience of their community.

In a fascinating but largely unsung slice of Western history, during the 1600s through roughly the mid-1700s there arose in nearly all the major Protestant churches of England a powerful movement celebrating the generous, relational quality of humans. Appealing to "the man of feeling," this perspective served as a much-needed counterbalance in a time when many considered humans to be hopelessly wicked. The famous seventeenth-century philosopher Thomas Hobbes, for example, saw people as by nature debased, capable of decency and kindness only when muscled into good behavior by frighteningly stern, powerful leaders. The "man of feeling" movement pushed back as well against the brittle notions of Puritans, who back then routinely insisted humans were cursed by depravity.

Identified by the somewhat chewy name of Latitudinarians, this enormous and enormously lighthearted group spread a message of exquisite comfort, extolling a natural impulse in humans to gain freedom through caring, tenderness, and charity. While the leadership of a great many churches resisted the Latitudinarians, fearing they were giving a green light to human passions, their messages of kindness and tolerance flooded the sanctuaries and lyceums of England for more than seventy years.

Latitudinarians called on people to extend their natural reservoirs of goodness to their fellow creatures—much, they said, as the goodness of God had been extended to all of creation. But more than this, they claimed the joys of being generous weren't limited to simply feeling good at having performed a specific act of charity. It was more fundamentally about the joy of experiencing the tender passions, natural to each of us, that spark such generosity in the first place. Here, then, was a great source of energy. Tenderness and sympathy were not weaknesses, but expressions of the unassailable kindness that would ultimately save the world.

These joyful passions, as the Bishop of Derry pointed out in a sermon in 1700, were not mortal dangers to be rooted out at all costs. They were the wellsprings of liberation. "'Tis our Passions and Affections that must do the work, for till they begin to move, our reason is but like a chariot when the wheels are off."

And then this extraordinary sentiment from a 1755 letter by a writer known today only as A.B.:

> Moral weeping is the sign of so noble a passion that it may be questioned whether those are properly men who never weep upon any occasion. They may pretend to be as heroical as they please, and pride themselves in a stoical insensibility; but this will never pass for virtue with the true judges of human nature. What can be more nobly human than to have a tender sentimental feeling of our own and other's [sic] misfortunes? This degree of sensibility every man ought to wish to have for his own sake, as it disposes him to, and renders him more capable of practising all the virtues that promote his own welfare and happiness.

William Sherlock, the dean of St. Paul's Cathedral and an early Latitudinarian, preached brilliantly about the value of a soft and tender mind—one that can feel the suffering of others, and suffer with them. "Nature prompts us to ease those sufferings which we feel," he said. And that natural impulse—what he called an "inward principle"—is more powerful than any external argument. "Sense and feeling is this principle," Sherlock suggested. Eight hundred years earlier, Aristotle claimed that humans were naturally "akin and friends to each other." The basis of that kinship, added the Latitudinarians, was to be found in inborn leanings that were comparable to "those instincts which are in brute creatures, of natural affection and care toward their young." We are by virtue of our own nature drawn together. And the best chance of happiness—not just in the future, but in the here and now—is to live each day as close as we can to that shining truth.

That these beliefs would have such enormous purchase is all the more extraordinary when you consider they took hold in a time of obsessive, hard-edged rationality. The powerful white men who ended up writing the history of this era gave little credence to passion and gentleness. And yet the profound generosity held in the messages of the Latitudinarians and their allies was like an artesian spring to the people of that country. It quenched the deep longing they had to see the world as a kind place—one that thrived best when humans directed careful, loving attention toward sustaining relationships.

While they probably wouldn't have described their philosophy in terms of efficiency, in a sense the Latitudinarians were promoting exactly that. A person kind to others, after all, tends to be more kind to herself. And that can in turn leave her more able to cultivate and apply her unique creative talents. Such a human community ends up embracing that essential circle of nature spoken of by Lao-tzu, taking from what is abundant and giving to that which is deficient. Living in this way lets us feel like we're coming home again, able at last to feel again the harmony, balance, and rhythm forever singing through our lives.

# After Disaster and Disruption: Nature Teaches Us the Fine Art of Rising Again

. . . It may, in fact, be necessary to encounter
defeats so we can know who the hell we are. What
can we overcome? What makes us stumble? And
fall? And miraculously rise? And go on?

—Maya Angelou

In the long, parched summer of 1988, not far from my home and across the wild heart of Yellowstone, there would come an alarming contagion of lightning strikes. And with the wheatgrass and bluestem paper-dry, with many of the trees themselves at moisture levels lower than kiln-dried lumber, the land began to burn. The wind blew, and small burns became great conflagrations. Flames heaved and swelled to more than a hundred feet high. The normally bright skies of summer were soon inked with smoke.

On August 20, a day firefighters came to call Black

Saturday, one of those burns, the Storm Creek Fire, thundered into the outskirts of the little town of Cooke City, Montana. Moving more than ten miles in three hours, it had built to such a headlong rush that it nearly singed the tails off a string of packhorses. The cowboy leading the string pulled them down the main street of town at a fast gallop, a curtain of smoke and flames behind them, producing a camera shot that made news editors and TV producers across the country swoon.

Many would later describe the Yellowstone fires of 1988 as the beginning of the modern era of wildfire. It wasn't that we were strangers to fire. In fact the West is a land thoroughly shaped and made healthy by the touch of flames. Yet for most of history the fires that held those flames were, more often than not, fairly modest. Early in the twentieth century, though, we decided that wildfire was the enemy. It was even portrayed now and then as a predator, this in a time when all of nature's predators—wolves and coyotes and mountain lions and hawks and owls—were thought evils to be slaughtered. For sixty years we put out every wildfire we could get our shovels on. And behind us, out of sight and out of mind, the forest floor became increasingly thatched with tumbled heaps of downed trees and branches—a so-called fuel load. One that in the past would've been cleaned away by those more modest burns.

The problem of heavy fuel loads in the forests has been greatly amplified by the growing effects of climate change: Severely diminishing the snowpack. Drying out the forests.

Locking us in prolonged droughts. Making the fires that spark from lightning strikes or careless humans bigger, hotter, more catastrophic. The ferocious Yellowstone fires of 1988 were a bridge linking what we'd done wrong in the woods with what we'd done wrong in the environment at large.

The following summer, 1989, nine months after the snows of autumn had lent their cold embrace to quiet that burned land, I set off for a five-hundred-mile walk around the eco-system. A good share of my route wound through vast ex-panses of blackened tree trunks—gaunt shadows of what were once hushed conifer forests. On some of the trees the charred bark had split open—signs of an especially hot burn—revealing slashes of smooth, alabaster-colored bare wood un-derneath. Without the millions of insects that lived in those trees, most of them having been lost in the flames, the wood-peckers and flickers and sapsuckers had flown to richer feeding grounds, leaving the woods without its drummers.

At the end of my second day on the trail I reached a high vantage point with a sweeping, ten-mile-long view of the landscape below. In every direction the world had been consumed by fire. Two-thousand-foot reaches of mountain stood hard and bare, their tattered, wind-sheared cloaks of timber roasted and crumpled on the stone.

A forest ultimately survives wildfire in part because of strat-egies it developed to withstand the blunt trauma. Ponderosa pine drops its lower branches as it matures—thus keeping flames from climbing branch by branch into the crown of the tree, like one would ascend a ladder. That tree also adapted to fires across thousands of years by growing extraordinarily thick bark, insulating the precious cambium layer from excessive heat. Besides that, a mature ponderosa forest spaces itself loosely, which both conserves water and limits the intensity of fires. Indeed it was these adaptations to fire that led early explorers of western North America to fall head over heels in love with the sun-dappled, parklike groves of the ponderosa—finding trees so widely spaced, rising out of ground so free of clutter, that it would've been possible to navigate them in horse-drawn wagons.

At the same time, when a major upheaval like wildfire moves through a landscape, we can predict how likely that system is to recover and thrive again by looking at what remains of it. Have the seeds necessary for sprouting new life been protected? Lodgepole pine, for example, produces cones that burst open to release their seeds only when touched by flame, thus becoming the first of the trees to reach down through the ashes and begin to grow. Is the soil stable and still fertile? Are bees and flies still in the area, thereby allowing the pollination of young plants? Is the aquifer still intact? Nature is all about keeping what's most essential from perishing, and then, with those essentials in hand, restoring vitality and stability as quickly as possible.

It's worth noting that in many cases life systems don't simply survive such upheaval—they *thrive*. Over many thousands of years, plants, insects, birds, and mammals evolved to take advantage of the bounty healthy fire leaves in its wake. Fire releases nutrients that have been bound up in trees as well as ground plants, feeding them back into the soil as ash. Which leaves the ground superrich. Despite my astonishment at the staggering visual impact of the Yellowstone fires, the skin of the landscape was, nine months later, already knee-deep in fresh runs of twisted-stalk and fireweed and spirea. And on ground that for nearly a century had been shaded by closed-canopy forests, grasses were exploding in the sun.

It wasn't just the quantity of vegetation that increased, but the quality. One study in Yellowstone the year after the fires found that fireweed growing in the burned areas was 30 percent higher in nutrients than the same plant growing on unburned land. And that had real significance for plant eaters. The elk of greater Yellowstone made a special point of dining on that postburn vegetation—an extraordinarily nourishing feast that helped them get healthy and strong in advance of the coming winter. In the following year beetles that dine on dead trees moved into the burned timber in droves, which in turn brought back those hard-drumming flickers and woodpeckers, all eager to dine on the bounty of insects.

It was in my long wanderings through Yellowstone in the years after the fire that I came to celebrate the small, early movements of restoration that I really hadn't paid all that much attention to before: a small pinch of whortleberry poking its head out of the scorched dirt; a lone aster calling out from the black of the burn. Nature in the wake of wild-fire shows itself as a process—sometimes fast and sometimes sluggish, yet infinitely persistent. Once I started really look-ing, I became able to see the influence of wildfire all around me—seeing the hand of flames in the very shape of trees, as well as in how those trees spaced themselves as they matured. It was, and would remain, a land created and re-created from the ashes.

This insistent flow of life—one that's especially well dis-played following a healthy fire—simply can't be derailed or denied. Imagine a fire burns to the ground a small patch of forest on an old abandoned farm, maybe somewhere in New England. The landowner hires sawyers and dozer operators to come in and remove the remains of all tree trunks and burned branches, more or less taking the area down to bare ground, fully exposed, almost like a plowed field. And then let's say that instead of getting replanted, the field gets aban-doned. The landowner walks away, doesn't give it another thought.

But for nature, "abandonment" is out of the question. In short order, that charred New England landscape will have lichen and fungi coming online, followed by a scattering of invasive weeds like thistle and burdock. Then will come chickweed and shrubs and tufts of grass. And all of that fol-lowed by saplings that progress into taller trees, many of them standing with their feet in thick patches of wildflowers, from violets to buttercups. In turn this fledgling woods brings in birds and small mammals. Indeed, it's been found that a single tree appearing in an otherwise unforested pasture can increase the bird diversity almost fortyfold. As for the creatures that were living under the forest before the fire came along—worms and beetles and bacteria—many of those would've survived the initial flames, continuing to drive the nutrition engine of the land by breaking down leaf and plant litter into the essential nutrients needed by those wildflowers and shrubs and saplings.

The miracle to keep in mind here is that the bits of nature that set up shop in the years after a major disruption will always lead to still more nature. And that nature will lead to still more. Even after descendants of all the old residents are in place, the system will keep expanding—taller trees, bigger mushrooms, thicker moss.

We know these fundamental things about life on Earth: Plants capture energy from the sun, using it for growth and reproduction. They then pass that energy along—to the deer that eat the grass, the squirrels that eat the nuts, the birds that eat the berries. Some, too, gets passed along as nutrients through the process of decay. And both during and after lots and lots of such passing along, what's left of that original sunlight energy finally gets released from the system—as heat, generated by the "work" of growth. What at first glance might have seemed like calamity, like an ending, will in fact reveal itself to be a highly coordinated, intensely robust, multifaceted burst of creation.

Following my five-hundred-mile walk around Yellowstone after the '88 fires, I sometimes found myself wondering what such an event might have to teach me about my own life. That thought hasn't gone away. Today, after all, we're living in a new age of wildfire, this one marked by so-called megafires—a term that refers to burns of more than one hundred thousand acres—which were once rare and are now common. Indeed since 2000, thirteen years have passed with at least a dozen such fires. Not just bigger burns, but far hotter. So hot that they often sterilize the soil, a condition that doesn't necessarily prevent the eventual rebirth of the forest, but can delay that recovery for a very long time.

And in wildfires I have found parallels to human life. The impulse to prevent even healthy wildfire has shown up in my own illusions of control. Because I grew up in a volatile home, I came into adulthood seeing any sign of tension as a danger sign. Even modest disagreements were hot sparks, and it was only prudent to put them out as fast as possible. But of course over time a staggering amount of debris builds up—feelings of not being heard or of being disrespected. That debris gets deep, choking the potential for new vitality in a given relationship. Suppress the small fires long and often enough, and one day you instead get a very big one.

But that's only the beginning of what fires are teaching me. I've also started trying to build clearer connections to what feels like my essential nature—to those core qualities that can withstand whatever challenges burn through. I look for what healthy "seeds" I'm holding right now that would be quick to start growing on the tail of trouble. I remember, too, that recovery from fire in nature is a communal act, a magnificent display of interdependence. With that, I'm reminded to take care of my relationships, nurture them, knowing that any one of us could draw on the others should disaster occur.

Arguably, the consequence of disruption is growth. Quite powerfully, forms of assistance like psychotherapy are designed to help us walk with eyes open into our annoyances—those modest burns of anxiety or depression that show up when life seems to be spinning out of control. And then to turn such disruption into opportunities to gain coping skills—to establish new flushes of mental, emotional, and spiritual liveliness. Disruptions can lead to healing. They can be what allows us to finally leave a toxic job or understand the true cost of staying in an unhealthy relationship.

But what about more serious, more catastrophic disruptions? Experiences of violence, or the death of a loved one? These are the equivalent of the Yellowstone wildfires—cusp events that greatly knock back the forest but stop just short of sterilizing the soil. They're the events in our lives that call most deeply for careful tending of the single blades of grass and sprigs of wildflowers that mark the opening acts of nature's return.

Our most distressing disruptions carry a risk that we'll end up stuck in fear, sunk into anger, or fiercely guarded. They may open doors to unacknowledged rage or to locked-away experiences of abuse or harassment. And in the same way megafires can rob the soil of its nutrients, the emotional threats following major trauma can potentially change the brain. Following a traumatic event, it's possible for our brains to become what some researchers call "bottom-heavy," meaning that the amygdala—that cautious, watchful part of the brain where fear responses are generated—becomes jumpy. At the same time, the parts of the brain that regulate emotions and allow clear thinking can become sluggish. Unprocessed traumatic events are likely to leave us anxious, depressed, unable to sleep—conditions that have serious health consequences of their own.

An ecosystem makes a multilayered response after a disturbance, when seeds and water and stable soil and sunlight come together to launch wave after wave, layer after layer, of new life, each building on the other. In the wake of emotional trauma, the first wave of recovery involves calming the fear center.

If trauma sufferers can early on just make the choice to move, to be physical for twenty or thirty minutes a day, it can help ease the hyperarousal states that follow tragic events, while at the same time releasing chemicals to help repair the nervous system. And mindfulness, from quiet meditation to being mindful of the food you choose to eat, is like gentle rain falling on burned soil.

Nature is a strong prompt for this kind of mindfulness.

This is why today an increasing number of returning combat veterans are choosing to begin their lives back home by first spending time in wilderness. And a growing number of groups, from Outward Bound to the Sierra Club to Veterans Expeditions, are reaching out to help make that happen. Besides providing the simple gift of quiet time to think, walking a wilderness trail causes the brain to produce powerful counterbalances to the toxic corticosteroids that occur during post-traumatic stress.

Out in nature, the world reaches out in every direction to create connections: from a group of trees sending nutrients through a fungal network to aid a sick maple, to wolves tearing open a carcass so that an old alpha male with worn-out teeth can feed himself, to Carole Noon's chimpanzees, comforting one another after the death of the woman who saved them.

And what's true for nature is true for you—you are, after all, a part of nature too. Having good, caring people nearby brings forward much of what's needed for healing. Some will be ones with whom early on you can tell your story, and then tell it again. People you can cry with. Others, including children, will be ones with whom you can gain tiny glimpses of the beauty that life "out there" still holds. Others will be there simply to give you the opportunity to open up to kindness, to be the recipient of a sacred giveaway. This, then, is how the heart reconciles itself, as poet Stanley Kunitz described it, to lives that must inevitably include a "feast of losses."

Research by Dr. Susan David of Harvard suggests that fully a third of us "either judge ourselves for having so-called 'bad emotions'" or actively try to push them aside. "Normal, natural emotions are now seen as good or bad." She goes on to explain that when we push away emotions like grief to embrace a false sense of positivity, we lose our ability to accept the world as it is, not as we wish it to be.

"Discomfort is the price of admission to a meaningful life."

We have the natural capacity to build more nourishing, empowering stories about our lives. To be kind to ourselves. And also, to build resilience through bonds with other people—with even a small cluster of caring friends and family who can help us get back on our feet in the wake of disruption.

In 2005 I experienced one of the greatest upheavals of my life, when my first wife, of twenty-five years, and I suffered a tragic canoeing accident in northern Ontario. Swept into a long run of ferocious rapids, the boat capsizing about a hundred yards in, I managed to escape with massive bruising and a couple of broken bones. For three days Jane was missing, rescue crews searching for her from sunup to sundown, by air and on the ground. And then a search dog, trained to pick up scent coming from under the dark, tea-colored river water, finally caught a sign. A high-ropes recovery operation was launched, and in midafternoon on a cool day in late May, fitful with rain, her body was gently pulled from that tumultuous water. I was nearby, at the search headquarters, and when the search commander told me the news I fell to my knees. And the long, often hopeless journey through grief began in earnest.

While on one level I welcomed the idea of friends and neighbors and relatives providing comfort—at times I craved it—the idea of those connections providing any kind of significant healing seemed far-fetched. Still, the gravity of that tragedy had thoroughly ejected me from life as I knew it; burned to the ground and helpless, I placed myself in their care. The memorial service was held on June 10 in what was then my hometown, in the Catholic church—the only place big enough to hold the event. I lurched through the side door of the church on my crutches, my brother at my side looking nervous, as if he thought I would stumble and fall at any moment. Inside the sanctuary, some six hundred people were waiting. Friends and family had gathered from all over the country.

I'd asked for room in the service for people to tell stories, and they came in good measure. Funny tales. Lovely ones, too. When it was my turn I explained how Jane often remarked that when it came her time to go, she hoped the end would come in a wild place, doing what she loved. And so it did. Somehow, though, I said, I imagined the end being decades away. Maybe with her as an old woman out on some last camping trip, snugged in a down bag, staring out the door flap of the tent into a sky riddled with stars.

The Red Lodge Fire Department, which included most

of Jane's fellow EMT and search-and-rescue workers, had parked their biggest fire truck outside the church, the ladder raised in tribute. Near the end of the service the dispatcher issued a so-called final page—an honor given to those who die in the line of duty, or who've made significant contributions to the community.

The radio crackled: "Red Lodge Fire Rescue, dispatch." Then a few long seconds of quiet, a gentle whoosh of static.

"This is a final page for Jane Ferguson. She died in the wilderness, doing what she loved. Her dedication and compassion for her fellow citizens will not soon be forgotten."

And then: "Dispatch clear." For about a minute afterward it was completely quiet; the whole place seemed about to collapse into sobs.

Looking back, I can tell you that the so-called fear center in my brain was on overload—great floods of troubled dreams and flashbacks of the wreck, as well as a compulsive loop where I kept imagining a dejected, despairing life that I was sure would plague me for the rest of my days. Yet at the same time I was learning—slightly, but deeply—how friends, even new friends, how even my two cats, could in a time like that shoulder a share of my burden, how they could begin to stem the bleeding of loneliness.

When my leg healed I started heading out into nature on long treks—officially to scatter Jane's ashes, as per her request, in her five favorite wilderness areas of the West. But unofficially, and largely unconsciously, because the community I also needed—much like Walt Whitman after his stroke—was one of creeks and forget-me-nots and deer and bears and mountain lions. Those places, too, teeming with all manner of life, would, like my friends and family, leave me feeling held, allowing me to feel at least a whisper of a sense that somewhere deep down I still belonged, and that such belonging might one day allow my grief to be just one facet of a life bigger than just that. From where I stand today I can see what happened. As those scattering journeys unfolded across the years following the accident, I was carried slowly but surely out of the little room where loneliness lives back into this wide world of embrace. A world of sunlight in my bones, of elk and wolves and tall pines and rivers in my blood.

My third scattering journey was to the flowing slickrock canyons of southern Utah. Back in her late teens, it was there, on an Outward Bound course, that Jane had finally turned the corner on what was a nearly fatal eating disorder she'd been struggling with for years. When she'd first decided where she wanted her ashes scattered, which she did some dozen years before her death, canyon country was high on her list.

I drove south from Montana, and by the time I reached the village of Caineville, Utah, the land had melted into the bare bones of existence: rusted waves of sandstone peeling away with every passing storm; deep blue sky, hot and thirsty and bright. Just off the highway, pocket meadows no bigger than backyard swimming pools nestled against sweeping arcs of sandstone, tossed with the red of firecracker penstemon, the burnt orange of globe mallow. Lone junipers hung by their toes to the walls of Slaughter Canyon.

I laid the brown pottery vase that held her ashes in the top of my day pack and started walking from a set of corrals near the Notom Road—heading west toward Sheets Gulch and the stark, fluted edges of the Waterpocket Fold. A cluster of cottonwoods was leafing out along the wash, dripping with that fleeting, electric green of April. The skies were mostly clear, though far to the west was a long train of dark clouds, dragging their tails along the tops of a red rock divide. Nature moves fast here, often violently, with storms entirely out of sight sending walls of water pushing down slot canyons, tearing boulders loose, and ravaging the cottonwoods. Yet another good reason to be fully anchored in the moment, paying attention.

Fluttering on the ground were clusters of Apache plume and rabbitbrush; along the damp edges of coulees, horsetail poked from the Earth looking like thin, bony stalks of asparagus. The magpies were out in force, rising and falling in ten- or twelve-foot dips, toying with the wind. I looked for just the right place, the right arroyo, the right butte, and in the end for unknown reasons found myself on top of a small rise at the eastern edge of Capitol Reef National Park. It was a view that whispered of a timescale so grand as to be inconceivable: old swamps in what is now the tumble of the Chinle Formation; massive desert dunes locked away in Navajo Sandstone; the hiss of shallow seas, now frozen in the layers of Mancos Shale.

The puffs of ash I spooned into the sky that afternoon held together for a long time, hanging in air that had all of a sudden turned windless, drifting slowly, slowly to the north against a reach of rust-colored sandstone. I placed the spoon and jar in the sand at my feet. Then I lowered my body to the ground, laid my cheek against a warm slab of rock. A lone, pumpkin-shaped cloud drifted overhead and then dissolved. A hummingbird flew by on her way to grab lunch from a patch of star lilies, passing so close to my head that I could hear the whir of her wings. The beauty of it all was impossible to miss. And for a precious few minutes there came a sense of putting the burden down. Like the hole in my life had gotten smaller, a smear of black in a bigger world of sky and slickrock and morning glories.

My recovery was a small natural system rebooting after a psychological wildfire of terrifying proportions. Like that patch of ground in New England, ravaged first by fire and then by bulldozer, laid bare and abandoned only to soon begin recovering layer by layer. In the end for me, too, life would yield still more life. More diversity in relationship and experience. More gratitude. More beauty. Like that New England landscape, I would in time be righted, friend by friend, plant by plant, bird by bird, until one day my ravaged heart and brain would return me to my own unfolding.

"Is the first note a part of the next note?" master cellist

Yo-Yo Ma once asked. "Or have you just moved from one infinite universe to another? The second note is always a revelation."

Not long ago, psychological researchers undertook a massive study involving some five thousand people. What they found is remarkable: While the level of happiness a person feels being connected to family, friends, or home is high, higher still are correlations between well-being and a person's connection to nature. Nature, for reasons we don't fully understand, appears to be uniquely slotted in our psyche—different from every other relationship we experience. Even more compelling, these researchers conclude that "nature relatedness" predicts happiness *regardless of other psychological factors.* In other words, in times of turmoil, those who have even a small slice of nature to lean on in their daily lives are more likely to prove resilient, emerging from difficult circumstances emotionally intact.

Wildfire, as well as floods and earthquakes and hurricanes, reminds us of what myth and story have revealed for thousands of years—that all life dances in a balance of creation and destruction, weaving and unraveling, with the new often arising with more strength and vigor than what came before. In fact, there's an old Taoist story that reveals well the seedbed of calm that lies beneath even the most challenging circumstances of our lives.

There was a humble farmer living in southern China, the story goes, and one day the latch on the stall of his most trusted workhorse came loose; quick on its feet, the horse ran out of the barn and disappeared into the countryside.

"Terrible luck!" his neighbors offered, shaking their heads with genuine sadness.

"Maybe. Maybe not," the old farmer said.

The very next day, not only did the horse come back, but he was attended by three wild horses. Incredibly, the four animals strolled into the corral next to the barn and began to feed on the hay.

"What a miracle!" said the neighbors. "A more wonderful thing could not have happened."

"Maybe," was all the farmer said.

The next day dawned clear and bright, and the farmer's son decided to take a chance at riding one of the new, untamed horses. No sooner had he climbed on the animal's back than it took off running and bucked against a rail of the corral, breaking the young man's leg.

Back came the neighbors, this time full of sympathy. "Such bad luck," they offered, sincerely sad for him.

"Maybe," said the farmer. "But maybe not."

The following day a team of brusque military recruiters came to the area, knocking on every door to secure young men for a war brewing in the hills far to the west. Seeing the farmer's son had a broken leg, the recruiters passed him by.

"Such fortune!" exclaimed the neighbors, astonished by how well things had worked out.

"Maybe," said the farmer. "Maybe not."

Much like the forest, we can and do rise from the ashes of even the most troubling events. True, in the wake of big trouble, most of us will have a hard time mustering the patient wisdom of that Chinese farmer. Yet we can lean on the comforting reassurance that as part of nature, with nimble brains and abiding instincts for connections honed over countless generations, we stand today in possession of a deep and steadfast resilience.

# Old Growth: The Planet's Elders Can Help Us Be Better at Life

The longer I live the more beautiful life becomes.

—Frank Lloyd Wright

As you read these words, somewhere in the warm, clear waters off the coast of Australia a mature bottlenose dolphin is swimming with her daughter. It's dinnertime. But Mom, instead of chasing down a fish in open waters like she usually does, swims over to a basket sponge growing on the ocean floor. In a deft move she breaks off a piece of the sponge, then fits it snugly over her beak, her rostrum. It's hard not to wonder what that curious, watchful youngster might be thinking about all this. *Are you going to eat that sponge? Are we playing?*

With the sponge secured on her beak, the older dolphin starts sweeping her head back and forth across the ocean floor. She's looking for bottom-dwelling fish like the sand perch, which hide themselves on the floor of the sea under layers of sand. As for the sponge stuck onto her rostrum, it allows her to clear away the sand without injuring herself on broken chunks of coral, or maybe even suffering the sting of another bottom-dweller, the scorpion fish. The extra work it takes to catch a fish like the sand perch is worth it because bottom-dwellers tend to be more fatty. And for a dolphin, more fatty means more nutritious.

Sure enough, after only about five minutes, a sand perch flushes. The fish dashes off for a few yards and then hesitates—waiting for a moment before burying itself in the sand again. In that brief pause the elder dolphin shakes off the sponge, surfaces for a breath, and then comes down and snags the sand perch before it can rebury itself. She then passes it to her daughter. And with that, the younger dolphin hasn't just gotten a good meal, but more important, she's learned a powerful hunting technique—one that years from now she'll pass along to her own offspring.

Meanwhile in another ocean, this one ten thousand miles to the northeast, orcas in the frigid waters of the Arctic are showing off some clever hunting tricks of their own. Having spotted a sea lion out on a small ice floe, three adult whales come together side by side, about fifty yards away from the ice. As if on cue, they quickly swim toward the sea lion in perfect unison, submerging together at the last minute, just feet from the edge of the floe. The result is dramatic. The diving bodies of the three whales create a line of big, fast-moving waves that roll across the ice floe, knocking the sea lion into the water. A young whale is nearby, watching all this unfold. And she doesn't forget.

And that's only one of many clever orca tricks a young whale will observe and learn. Less than a mile away a group of five orcas is coordinating something even more complicated. Two whales swim out a few hundred yards, turn around, then start slapping their massive tails against the water. The sound of that slapping, which can be heard hundreds of yards below the surface, alerts nearby fish to the whales' presence; in no time those fish start moving in a line away from the sound, unaware that they're essentially being herded by the whales. Meanwhile in the distance the other three whales have gathered in the deep. At exactly the right moment the group blows a massive net of air bubbles that rises to the surface and for a short while actually traps the fish, holding them just long enough for other orcas to swim over and feed.

All around the world, on every continent and in every sea, wisdom is flowing from mature adults to the less experienced. Meerkats are teaching their young how to handle the scorpions they find so tasty, without getting stung. Chimpanzee leaders are soothing chimps who just lost fiery arguments with their cousins, sitting with them as their anger settles. Wolf leaders are guiding less experienced members of their packs across miles of rugged mountainscape— and when they finally find elk, showing them how to hunt in a way that reduces the risk of injury or death from being kicked.

Older orangutans in Sumatra are teaching their offspring the complicated work of building a proper sleeping nest in the branches of the trees—lessons that may stretch across four or five years. Even ants can be thought of as teachers. Somewhere on a grassy patch of ground in France, a female rock ant has located a valuable food source about ten yards from the nest. On returning to the nest she gets the attention of a young fellow ant, and, in service of sharing the goodies, begins leading the younger ant back to the food source. As the pair travels, the leader pauses every so often for the following ant to get its bearings—to take in vertical landmarks; in this way the student memorizes the route, which will allow it to return to the food source later, on its own. At the end of each of these brief pauses, when the follower is fully oriented, it taps the leader on the back and the pair continues on its way.

In nature, the strongest creature cultures are those that balance the energy and strength of youth with the experience of those who've been in the game of life for many years. The more social the species, the more valuable are the mature leaders—those who can apply their experience to a wide range of needs in their communities. This, then, is leadership based not so much on dominance, but on reputation and relational skill. On standing.

Like elephant leaders. Those elephants who stand at the top of herd culture have decades of experience not only learning the nuts and bolts of survival—things like how to find water in a drought—but also negotiating the personality and behavioral issues among the other elephants. It takes years of paying close attention for an elephant to learn what to do when two individuals in the herd, or even two groups, aren't getting along. Likewise with knowing how best to direct younger, more aggressive herd members during an encounter with a pride of lions, or even with other elephants.

Not surprisingly, the absence of such reliable wisdom can lead to wounds in the community—wounds that include everything from unfounded anxiety to neglectful parenting to extreme aggression. A number of years back, a group of male elephants from South Africa, orphaned when they were young, grew up to be extraordinarily aggressive, getting into fights they had no business being in—and incredibly, over a period of ten years, killing more than a hundred rhinoceros.

Years later, a brilliant piece of research by psychologists at the University of Sussex revealed how the loss of elder wisdom among elephants could cause deep misfirings in the young. The research took place with another herd of males in South Africa's Pilanesberg National Park, which in the 1980s and '90s had become orphans due to a human culling of the herd. The scientists wanted to compare the social skills of that group to those of a group of elephants in southern Kenya's Amboseli National Park, where the young had retained all of their familial connections.

Each group of elephants was exposed to audio recordings of different vocal calls—their own, as well as those of elephants they didn't know. The calls from strangers were from elephants that varied widely in both age and size, which basically means that each voice would've conveyed a different social ranking in the herd. Researchers with video equipment recorded the herds' reactions, taking special note of bunching behavior, which elephants display under perceived threats, as well as the extent of their smelling and listening behaviors.

The elephants from Amboseli, whose families were intact, scored high on discerning familiar voices from those of strangers, keying in on what could've been actual danger from unfamiliar animals. They were also able to distinguish stranger elephants of different ages, appropriately showing more caution—and more defensive responses—when they heard the voices of older animals. This ability alone—to figure out the status of a stranger—is critical for creatures who live in complicated social networks, coming into contact as they do with hundreds or even thousands of other individuals. It's essential for keeping conflicts to a minimum.

Yet the orphaned elephants of Pilanesberg had no such abilities. They failed to discern friend from potential foe, as well as older leaders from younger, less dominant animals—deficits that could have deadly consequences in the wild.

What this research couldn't yet tell us is whether such deficiencies, and the considerable anxiety they cause, had left the Pilanesberg elephants unable to experience the contentment of their own group. What might go missing from such extraordinarily intelligent animals' lives when there are no elders to teach them whom and how to trust? Have the uncertainties within the Pilanesberg herd left them on most days in a kind of tense vigilance, prone to fear and aggression? Could such a state have long-term effects on their health, as it does in humans? Is it possible such anxiety might even alter the brain chemistry of their offspring, much like epigenetics is showing that human trauma can be passed from one generation to the next?

Nature works by giving the less experienced members of a community the chance to freely tap into the wisdom of their elders—learning from them essential knowledge about how to make a living, how to stay safe, how to move from one place to another. Even what to do with anger or grief. This transferring of essential knowledge is what allows the community to stay strong. And in the end, it's what gives the entire species a good shot at building the fitness they'll need for long-term survival.

It's pretty easy to appreciate the value of elders in the animal world. But as it turns out, maturity can be equally valuable in a forest. While older trees may not teach in the way we understand that term, they do share and communicate, and those actions are in full service of nurturing the community.

To get a better look at such elder sharing, we might make a morning visit to the magnificent coastal redwoods of Northern California. The shade would be deep and soft, with patches of low fog kissing the tops of the oldest, tallest trees. Starting as a seed only an eighth of an inch across, these redwood trees have grown to staggering proportions—more than thirty stories tall, and weighing in at a whopping six thousand tons. Having sprouted in the earliest years of Christianity, they're among the longest living life-forms on the planet.

Though once fairly widespread, today coastal redwoods live only at the very western edge of the North American continent. But even here the moisture falling from rain or snow is only half what they need to live. To compensate for that, the redwoods have learned to capture the tiny drops of moisture that regularly roll onto this part of the continent as fog. And it's often the tallest of these trees, the oldest, that manage to snag the most fog from the low, drifting clouds. That keeps them strong and healthy, able to feed and stabilize the entire forest, in the process creating ideal conditions for young redwood seedlings, many of which are just beginning their long, patient reach for the sky.

You may recall the short walk from my childhood home in South Bend, Indiana, into the oak forest near the St. Joseph River. Just as those robust trees are communicating with one another through astonishingly complex and efficient fungal networks, so, too, are these redwoods. Indeed, every time we take a step on the soft, needle-strewn ground under these giants, directly beneath that single footprint could be as much as ten miles of tiny fungal strands woven through the soil.

The healthy elders of this redwood forest are consistent,

generous users of this network. In a given year the big trees will send countless special deliveries through the fungal web—many bound for younger trees that are not only more vulnerable to disease but, because of their size and growing conditions, less able to muster the carbon they need to grow. In some cases elder trees may even use the web to prompt young saplings to activate useful genetic traits, such as a heightened resistance to drought.

Several years ago University of British Columbia forest ecology professor Suzanne Simard, a pioneering researcher of these fungal, or mycelium, networks, posed an intriguing question. While it's clear that trees communicate and exchange goods across species, could it also be that an elder tree is especially prone to tending to her own young family members? Simard's research suggests that's exactly what's going on. The biggest, most vibrant fungal networks are the ones between elders and their young relatives. What's more, being able to sense their young through such connections, older trees may even scale back their own root structure to give the saplings more room to grow.

Simard also says that if an elder tree is sick or dying she'll send extra doses of her own carbon to young relatives and may at the same time also help stimulate defense mechanisms in those youngsters. This gift of extra carbon and increased disease resistance gives the young trees a boost, preparing them to better meet the stresses they'll surely face across their long lives. So important are such actions by the grandparent trees, Simard explains, that the little trees under their canopies survive at a rate three or even four times what less-connected seedlings can manage.

"Trees talk," Simard says. "Through back and forth conversations, they increase the resilience of the whole community." And as often as not, the best, most helpful of that talk comes from the elders.

Lately these old trees have also been talking to us humans. Newly developed methods for taking wood-core samples from redwoods (without harming the trees) are allowing climatologists to figure out how much rain and fog occurred in a given summer in the past. Eventually they may be able to assemble records stretching back a thousand years. What's intriguing about that is that the presence of fog is closely related to the surface temperatures of the ocean. If we can tell how much fog there was in any given summer from years past, we'll gain a deeper understanding of long-term natural ocean current patterns. And that, in turn, will likely allow us to better understand the extent of human-caused climate change.

Many indigenous cultures around the world have referred to trees as "keepers of the stories." They have so much to tell us. Not just about things that happened in the past, like prolonged droughts, or the rise and fall of so-called little ice ages, when the climate went cool for a time. The oldest of the trees also have much to say about growing strong. About how a being thrives in the throes of floods and wildfires and roaring winds.

Here in the human world we tend to turn to parents, aunts, uncles, and teachers at school, all of whom spend a fair amount of time helping us learn how to navigate daily life. They teach us about work, about responsibility, about getting along with others. They try to send us off into adulthood with the kinds of skills we'll need to prosper in the big, wide world.

But beyond learning how to do math and drive a car and invest money and handle a prickly co-worker, we have been given a high level of consciousness by nature—a capacity for reflection and deliberation and empathy that leaves us with the challenge of navigating not just an outer world, but also a vast interior world. And being able to figure out that inner-world journey is firmly linked to some really practical concerns, like personal health and even longevity. For this human passage, nature asks us to build emotional resiliency, to learn how to get our feet back under us after heartbreak and setback. It asks us to develop, then actually trust, our intuition. To move forward in the face of fear. To figure out how to swim with depression without drowning.

As is true for many other animal species, for humans much can be learned from those among us who have lots of years under their belts. But hold it, you might be thinking. Your world is completely different from the one that people the age of your parents or grandparents lived in when they were young. That's true. But many of those differences re-side in nothing more than the trappings of technology. And even technology, to the extent it can overwhelm and over-distract and overisolate, will land you in dark places not altogether unlike those navigated by people who came long before you. What makes up the essence of who we are as human beings is at its core quite stable. And with no help from those who've gone before we can end up like those orphaned elephants of Pilanesberg National Park: strug-gling, clueless, feeling anxious and alone.

Back in the late 1990s I took that two-month-long, four-thousand-mile blue-highway journey in my old Chevy van—the one that early on landed me at artist Frederic Church's spectacular home on the Hudson River. From there I traveled to Maine and then down to the Carolinas, up to the North Woods, and finally across the Great Plains back home to the Northern Rockies. Along the way I'd had dozens of conversations with people, many of them older, who'd somehow figured out how to keep the lessons of nature strong in their lives. The project came at a rough time in my own life. And because of that I tried to make the most of every encounter, working to really understand how the outdoors had brought these elders peace, had shored up their resilience. I was headed north out of the hollers of East Tennessee and had one more stop to make in the Midwest before heading on to the North Woods. It was in the heart of my old hometown of South Bend, at a small white house on the very street where I grew up. Right next door to my own childhood home, in fact—the two little houses split by a ten-foot-wide ribbon of grass.

I knock on the front door, and after a few seconds hear rustling through the open windows. Finally, the door opens and a beautiful old woman, ninety-three years old and still topping three hundred pounds, stands leaning on her cane, smiling, welcoming me in with a sweep of her big brown arm.

"I'm washing my walls," Pearl tells me, maybe in response to the doubtful look she sees on my face at spotting the ladder standing in the archway between her tiny living room and kitchen. I insist on helping her, but she'll have none of it. Tells me that if she doesn't keep moving, she'll never be able to get going again.

"Besides," she huffs, all the while smiling, "I want it done right!"

We take a seat on her glassed-in porch, where we can look up and down a street that, save for a few old trees that have disappeared, has changed little. She tells me that tomorrow she's driving up to Michigan to help her niece can fifty quarts of tomatoes.

Pearl was next-door neighbor to my family for thirty-five years. Maybe because she and her husband didn't have kids of their own, she took my brother and me on with a love that I suspect even blood relations rarely see—unconditional, a bright and bottomless well. Sitting here with her now I'm thinking of how she'd have me over to help make cookies and how the flour and sugar would fly all over the place, just like the grease did when she fixed her famous fried potatoes in a big black iron skillet, as if food was somehow less of a celebration if there wasn't some kind of mess left in its wake. Given that I grew up in a place where orderliness was nearly a religion, her willingness to make messes was all by itself reason for me to love her. "You and Jim was always good boys," she used to tell us. She tells me again today. She believes it so completely, says it with so much heart, that even though I know better I can't help but think it might be true.

Pearl grew up just to the north, in Buchanan, Michigan, in hard times, so in love with the outdoors that by the time she was nine she was repeatedly running away from home with her cane pole, making for her favorite lake so she could forget the world and just fish. No matter the punishment her parents doled out, it never kept her away from that lake for long. Later, besides making regular hunting trips to Michigan, while she was in her forties she and her husband bought a tiny cottage on Lake Wawasee. From then on, all summer and well into fall they spent every day off from their factory jobs climbing into their respective boats and motoring out to spend several hours with fishing lines in the water. For a few years my brother and I would stay with them for a week in July or August, both of us going out every day at five thirty, Jim with Merle, and me with Pearl.

I didn't even like to fish all that much back then. But I loved to fish with her.

Maybe it was payment from God for her good heart, or some kind of compensation for the hard life she'd lived, but Pearl had an astonishing way of reeling in fish with a kind of success virtually unknown outside biblical miracles. She was a big woman even then, and the sight of her in her torn, worm-smudged cotton dress jumping up in that tiny boat to wrestle some big bass or bluegill or perch from one or more of the three poles she kept going at all times—well, for a ten-year-old it was like sitting at the feet of a shaman gyrating for rain, and always getting it. A holy frenzy of baiting and casting and landing.

To this day watching Pearl on that lake remains among the most amazing, illogical things I've ever witnessed. Sometimes, as if magnetized, people in other boats would edge as close to us as they thought proper. They'd bob about for an hour or two and catch absolutely nothing before finally motoring away in a huff. One time a little boy about my age was in a boat nearby. Like so many of the others, he was having no luck himself, but at the same time having to suffer watching one of Pearl's incredible harvests less than two hundred feet away. Finally, he looked up to see one of her fish breaking the line and getting free before she could land it. "Good!" the boy yelled. Made Pearl laugh so hard she couldn't sit up straight.

"Them was the happiest times of my life," she says.

I tell Pearl about my journey, about the birch-bark wig-wam I helped build with a Penobscot elder, about the beautiful paintings I'd seen in the mansion on the Hudson River, about the moose in Maine and even about the moonshine I drank with an old farmer in the hills of Tennessee. I tell her that from here I'll be going up north to places she used to go on hunting trips, following some of the same roads she took, roads my family followed, too, on a handful of precious vacations to the North Woods.

Most of all, though, I tell her I wish she could climb in the van and go with me. So she could show me the fields where she picked strawberries at fifty cents a day, or point out the woods behind her brother's place where she used to gather mushrooms. How I'd love it if we could maybe find that lake she kept running away to as a kid, or another one not far away, where she says she used to sit on a log at the shore with her sister-in-law just to listen to the singing of the birds.

Before I leave, she heads into the basement, then comes up ten minutes later with a box full of homemade relish, grape jam, canned beets and butter beans, a package of pork chops from her niece Margaret, four boxes of vanilla pudding mix, and a half bag of strawberry Newtons. Provisioning me for the trip north. "Now, listen," she says over my objections, sounding stern, "I've got more than I can ever eat down there. You just take this. You'll need it."

Lastly, she goes into a corner of her kitchen, where beside the refrigerator she's propped up her 1930s bamboo fly rod and reel. Taking careful hold of it, she passes it to me with a grin.

"Here," she says. "This is for you."

Pearl wasn't religious in the traditional sense, preferring to do her visiting with the Creator while babysitting bobbers in the middle of Lake Wawasee. But she'd shored up her life with two strong notions—ones she handed off quietly to me. The first was that fear has little room to grow in those who simply keep going, one day at a time. Second, that such persistence was fueled by gratitude. Clearly she'd managed all of it—both the persistence and the gratitude— through big, regular doses of nature.

The challenges Pearl faced—deep poverty, but also physical abuse—had caused her to pay very close attention to the world around her. Attention as a survival skill. But she learned to also turn that attentiveness to nature, feeding herself with the calm and beauty she found there, which helped her keep going. At the same time, at least from what I could piece together on that summer day sitting with her on her porch, her abundant gratitude came in part from having known nature as a source of unbridled abundance. By then she'd spent ninety years watching glorious vegetable gardens springing each year from the dirt in her or her family's backyard. And even more powerful were those long, sweet summers on Lake Wawasee when fish were all but jumping into the boat to feed her. I think Pearl was so generous, because nature taught her that was a good way to be.

A hundred miles north of South Bend I find myself still looking back at Pearl's fly rod. And at the box of food sitting on the floor, wishing she really had come along so that I could pull over in some patch of woods and cook up a big meal for her like she used to do for me. I'd turn the burner on the camp stove up real high, toss in the pork chops, and just let the grease fly.

Pearl died quietly at 102. But she left a light burning in me. Pearl helped me not give in to the temptation to build walls. To not get too fond of my own drama. And to practice always the art of looking in, by looking out.

The lessons in this book are in part about health. Wisdom. Gratitude and contentment. It's true that we humans have a curious capacity for too much consumption, too much striking out at others in grandly imagined fits of survival. But nature has given us a miraculous capacity to learn what makes things in the world strong and graceful and full of vitality, and then steward those same things in ourselves. Increasing the possibility of our saving the world by taking clues from the very things we're trying to save.

We have the great natural gift of being one of the species capable of learning across an entire lifetime. That can make elderhood especially powerful. The emerging brain science of neuroplasticity—which focuses on the brain's ability to reorganize itself—has turned to rubble old thinking about the limits of learning. For many decades the brain was thought to be fully formed and fixed by the time we reached eighteen, then starting to degenerate by age forty-five. Now we know that learning and changing our behaviors can happen throughout our lives.

Everyone starts out with a brain that has plenty of "plastic potential." But what can turn us into brittle, rigid caricatures of our original selves is the amount of nearly constant, narrowly focused repetition in our lives: the same commute, the same general tasks during our workday, the same television shows at night. This is why nature, with its boundless mystery and bottomless unpredictability, is so remarkably good for your brain. And it's why you stand to learn the lessons of this book most deeply when you give yourself the gift of actually connecting to the natural world.

What you can end up creating is a wonderfully positive feedback loop. Your brain becomes more active, more nimble, out in nature. And that more nimble brain, in turn, can change profoundly what you're able to see and experience in the world at large. As neuroscientist Anil Seth at the University of Sussex puts it, "We don't just passively receive the world. We actively generate it." And when raw material for generating our understanding of life comes from nature, we have the potential for lasting vitality. The natural world all around us, it turns out, is one of the best ways of taking care of the world inside, priming in each of us the ability to be not only at home with ourselves, but also a precious resource for everyone around us.

The wisdom of elders can help us all begin to look at the

world around us with eyes and hearts that see the essential power of interdependence and diversity. It can teach us the graceful efficiency of being better able to make right choices in ever-changing circumstances. Guided by those who have journeyed across the decades, we can walk into both the natural world and our daily lives without needing to know everything, able to welcome the mystery of it all. We can learn how to come through the fire, and from that burning become even stronger.

Elderhood expresses itself in humans and elephants and chimpanzees and wolves, and so many others, too, as a time of life resting in the power of relationship. Of kinship. Notably, across most of human society the idea of kinship isn't limited to blood relationships. It has to do as well with bonds formed with the people who simply share our lives—those with whom we eat, work, cry, raise our children. In some cultures, this sense of community even includes people who died long ago; for the people of these cultures, a gratitude arises from the fact that no matter their own accomplishments, they stand fully and forever on the shoulders of those who went before.

Among the Ku Waru people of New Guinea, kinship arises by way of something called *kaopong,* which is said to be a substance originating in the soil. And while a father's sperm and mother's milk contain *kaopong,* so do the locally grown sweet potatoes and the pigs being raised in the village. Just the act of sharing those foods makes people kin, no less so than a child is kin to the mother who gave him birth.

Our task now, with the help of nature's master lessons, and with the help, too, of the wisest human elders among us, is to grow our notions of kinship. To go beyond blood ties and the commons of place, beyond species even, to launch a sharing both humble and strong enough to honor the Earth and all the life it holds.

At which point we may find ourselves again cocking our heads to better hear the ringing of birds and the muttering of frogs. Feeling the reliable wash of gratitude that comes from knowing that whatever else may be happening, we're still here, wolves and lions and hawks and humans, all of us in our own way rising in the morning, faces to the sun.

# Nature's Beauty Holds Every Lesson

Beauty will save the world.
—Fyodor Dostoyevsky

We're ready now. Able with every new day to take small steps toward a more expansive view of nature. One resting in the wonder of our scientific discoveries, in the comfort of relationship and interdependence, and finally, in the mystery of all we don't know. A world marvelously untidy.

One autumn, hearing that a special guest was coming the next day for a visit to the temple, the gardener gets up long before dawn to set about trimming every branch, pruning every shrub, raking all the fallen leaves, and clipping every blade of grass, until finally, near exhaustion, he settles back to admire his handiwork. Now, as it happens, at that very moment an old Zen priest walks by the garden wall and stops to regard the gardener. Grateful for the chance to share his accomplishments, the gardener invites the priest in for a closer look.

"What do you think?" the gardener says, beaming. "Isn't it glorious?"

"Nearly so," says the old man, walking over to one of the trees at the center of the garden, where he puts his hands on the trunk and shakes it furiously, letting loose a tumble of dried leaves and small branches onto the clipped grass lawn and freshly raked pathways.

"Now," says the priest. "Now, yes, your garden is truly beautiful."

Beauty. Simple and uncluttered. And at the same time, shifting, messy, chaotic.

As Harvard aesthetics professor Elaine Scarry points out, while there are plenty of things in life that make us feel blissful, and plenty of others that make us feel marginalized, beauty has the extraordinary capacity to do both at once. Each of us makes frequent free falls into the mistake of thinking we're at the center of our world. Beauty stands by ever ready to relieve us of that burdensome illusion.

"Beauty not only puts us on the sidelines," Scarry says. It makes us very happy to be there.

This sense of being on the sidelines and yet feeling joyful about it is precisely how the beauty of the natural world allows us to feel suddenly and essentially connected to it. The beauty of nature isn't merely a launchpad for ruminating about our own lives, but rather a gentle nudge to get us happily out of our self-centeredness and into the wonder of being in and of it all. When the beauty of the natural world arrests your attention, the walls of an anxious psyche recede—even when those walls are made of the hard, thick stones of loss and sadness. Many philosophers over thousands of years have maintained that the experience of beauty is one that calls us to heal the injuries of the world. Quite so. But it's also a call to heal what's been injured, what's been lost from awareness inside each of us.

When it comes to beauty, you know what you know. You feel what you feel. Some of our most powerful early experiences as children rest in the delights of perceiving shape and symmetry in nature—the sweep and color of wings, the patterns of leaves, concentric circles from stones tossed into a pond, petals and rays spilling out in brightly colored rings from the centers of the flowers in the window box. For some, these graceful facets of the natural world become a source of creative energy for the rest of their lives. Like Isadora Duncan. As a child, the celebrated choreographer was drawn again and again to the smooth, fluid rhythms of ocean waves. She would later go on to inten-

tionally integrate those wave movements into her early bal-
lets, creating what we now know as modern dance.

Beauty carries us. And there's even encouraging evidence that our sense of what's actually beautiful is expanding. From about AD 500 through the 1500s Christian
leaders routinely cast aspersions on the irregular landforms
of the world. Rugged coastlines, the twisted folds of mountain heights, were a travesty. Insults to the state of perfection
they claimed existed shortly after creation, when the Earth
was round and smooth and sweet with order. When it was
a "mundane egg." Crooked, asymmetrical natural features,
these clerics said, appeared after the biblical floodwaters, put
here by God to remind us of our evil ways.

All that began to change in the seventeenth century,
when curiosity about "irregular" landforms took flight.
Landscape artists began revealing nature not as the hiding
place of the devil but as a source of spiritual rejuvenation.
Western poets and writers became willing to tie their own
lyrical notions of the Earth to emerging fields of science. As
noted earlier, Galileo's telescope alone created a tremendous
sense of possibility; a "psychology of infinity," as poet and
philosopher Henry More described it.

Following suit, philosopher Anthony Ashley-Cooper,
the Third Earl of Shaftesbury, began seeing magnificence
even in the most barren regions of the world. Vast deserts
that at first may seem ghastly, he said, weren't without a
peculiar beauty.

In the warmth of a June afternoon in the late 1990s, I pulled off Highway 53 in northern Minnesota and made my way into the streets of Duluth and then up to the modest lakeside cottage of an Ojibwa elder. It was a week exactly since I'd been in South Bend with my ninety-three-year-old former neighbor Pearl, still on that four-thousand-mile drive across the country to learn from people who'd never lost touch with nature. Several Native Americans I'd been talking with in Minnesota told me about this particular Ojibwa woman, a storyteller. I'd called her on the phone to ask if she'd be willing to talk with me.

"Yes," she said quietly. "But I only know what I know."

The day we met was crisp and sun-drenched. Even now I can recall the smell of Lake Superior wafting in through the open windows of her kitchen, where we sat at a worn oak table washed in summer light. At one point, after we'd been talking for an hour or so, she laid her brown arms on the table, palms up, and began telling me a tale that perhaps more than any other has changed my life.

It was a long time ago in the land of trees. Spirit Woman had given birth to human twins. Now, as it happened, it fell to the animal people to care for these babies, and they were committed to the task—doting on them, eager to meet their every need. Bear warmed them through the wee hours by hugging them to her hairy chest. Then each morning at dawn, Beaver came along, taking the babies from Bear and carrying them to the shore of a nearby lake, where she dipped them in the water and then set them out in the meadow in the sun to dry.

Then it was Dog's turn. Dog took his job more seriously than anyone. When flies came along and pestered the babies, Dog snapped at them to chase them away. When the twins were cranky, out of sorts with colic, he nuzzled their bellies with his cold, wet nose and made them laugh. If that didn't work he jumped into the air and did all manner of clever tricks. Deer gave them milk throughout the day. At night the birds sang them to sleep.

But something wasn't right. And one morning Bear got up the courage to say something about it. "We feed them and care for them like our own," she said. "But still they don't stand. They don't run and play." Everyone knew exactly what she was talking about. "Okay," said Dog, already making a plan. "Nanabush, the son of the West Wind, is coming tomorrow. He's smart. He'll know what to do."

Sure enough, the next day Nanabush did come. Because Nanabush always comes when the animal people need him. He studied the babies out in the meadow, all the while listening, nodding as the animal people explained the problem. First of all, he finally told then, you've done a good job taking care of these human babies.

"I think maybe you did too good of a job. The young of any creature doesn't grow by having everything done for them. They grow by reaching, by struggling for what they want."

But as smart as Nanabush was, he was clueless about how to fix it. So as he'd done countless times in the past, he readied himself for a long journey west, to a certain high peak he knew about—maybe it was right here in the Beartooths—to ask the Great Spirit what to do.

Nanabush left the land of the trees and began the long trek across the prairie, reaching that certain mountain after weeks of hard travel. With no small effort he climbed to the summit, and there summoned Great Spirit. And Great Spirit came. Because Great Spirit always comes when Nanabush calls. After explaining the predicament, Nanabush was told to start scouring the summit of that great mountain for a certain kind of colorful, sparkling stone. "Gather every one of them into a big pile, right here," Great Spirit said. It was a huge job. But Nanabush had been around long enough to know there was no use trying to bargain for something easier. He started collecting. Day after day after day, until finally there was an enormous pile made up of every last one of the colored stones.

But what was he supposed to do next? Hour after hour he sat there hoping for some further instruction from Great Spirit. But no word came. Finally, out of boredom, Nanabush began tossing the stones into the air, first one at a time, then big handfuls. He invented games. He learned to juggle. Then one morning, as the sun was poking above the east horizon, he grabbed a big handful of the stones and tossed them high into the air. Only this time they didn't come down again. This time they changed, turning from stones into the most beautiful winged creatures Nanabush had ever seen. They were the world's first butterflies.

Now he knew what he needed to do. He worked his way down the mountain and began the long trip back across the prairie, the whole time surrounded by a flashing, fluttering blanket of butterflies. When he finally got back to the land of the trees, back to the babies, the twins looked up from the grass and were overjoyed. Their arms went up toward the sky, and they were trying their best to catch the butterflies in their chubby hands. Of course that's no way to catch a butterfly. Pretty soon they started crawling after them. A few more weeks passed and they were on their feet, still reaching, still trying. In time they were walking. And then not so long after that they were running through the woods and across the meadows, trying to catch even one of those beautiful winged creatures.

And that, say the Ojibwa, is how butterflies taught children to walk.

I thanked her. We said little else, but as I got up to leave, the elder put her hand on my shoulder. She had an instruction. If I told that story, she said, and she hoped I would, I needed to understand something. I must realize that her people didn't keep that tale alive because they need to be reminded not to give their children everything they want.

"We know that," she said. "Instead, we use that story when we get stuck. When we fall into sadness or anger or lose hope. The story tells us to first heal our relationship with beauty. Beauty will help us start moving again."

I went home to Montana. I wrote. I walked in the mountains and thought about the butterfly story. Years passed, and one day I found myself in the gracefully sinuous canyon country of the Southwest, there to chronicle a compassionate wilderness therapy program for so-called at-risk teens.

One of my first nights with the kids—eight girls ranging

from fourteen to eighteen—we sat around the campfire; having figured out I was a writer, they asked if I knew any stories. The question caught me off guard. I ran one tale after another through my head, worried that what I knew might seem bland or boring to girls who'd been ravaged by heroin and depression and abuse. But then I thought of those butterflies. I took a breath and began to speak. They listened, intently. Leaning closer and closer toward the fire. At the end they smiled and nodded quietly. Several were crying.

I watched and listened and learned from those young people that spring. And I went on to follow a dozen of them for a year after they left the wilds, hearing about how their healing was holding up back in the world at large.

One of the girls I checked in with regularly was Alexis. Like quite a few others, she was from an upper-middle-class family, having arrived in Utah at sixteen with her life in tatters from a three-year-long addiction to heroin. She was a veteran of every kind of intervention imaginable, from twenty-eight-day lockdowns at various resident drug addiction facilities to an endless chain of individual and group therapy sessions, even a couple of short but scary visits to the Anaheim jail. Alexis and I met on the slickrock. She was furious. Bitter about having ended up in the godforsaken middle of nowhere.

A year after she completed the wilderness program, we were talking by telephone.

"Here's the thing," she started. "The wilderness ruined

my high. A couple months after I got home I tried junk again. But I stopped. Walked away from it. I knew too much."

Over the next half hour or so she offered some thoughts about what happened out there to make such change possible.

"It was the first time I've ever known beauty," Alexis told me at one point. "Beauty so deep it almost hurt."

She was hardly the only one to mention that particular gift of the wilds—this knowing beauty for the first time. And despite what the Ojibwa elder had told me all those years before, that nature's beauty would help humans find their way, that it would help us get moving again, I was frankly astonished that it could be such a powerful healer of the human spirit. I told Alexis it seemed sad that anyone could go sixteen years without ever knowing such things.

The phone was quiet.

"I should change that," she finally said. "I did know beauty once." Then she went on to describe how at seven she got to spend a week at her uncle's cabin near Big Bear Lake, in the Sierras. Wading in creeks, picking flowers for the table. But mostly she told of going with her uncle to the top of Castle Rock. Up there, by chance she caught sight of a hawk flying close by and at eye level, so near she could hear the sound of its wings stroking the air. She'd looked up at her uncle and told him she wanted to have wings, to fly like that. He said it was a very good dream, that she should try her best to hold on to it.

"Pretty soon after that my parents divorced. In the middle of that my brother was killed by a drunk driver. Everything fell apart."

Right now, some twenty years after I walked for the last time with her through the sage and rabbitbrush of southern Utah, Alexis is a pediatric nurse with two kids of her own. When I last spoke with her she was still adamant that her time in the wilderness was the most important experience of her life. And curiously, the reasons she gave for feeling that way hadn't changed: Among other things, it was where beauty had reached out and taken her hand. Where she'd finally taken her place as part of the bigger world—out there in the aspen woods high on Boulder Mountain, and in dozens of secret draws and quiet slot canyons in the Red Desert.

"Last summer we took our two girls to Castle Rock," she told me. "They were as excited as I was when I was a kid. They loved it. Maybe they'll reach back and grab that feeling when they get older, when they really need it."

Nature attracts us, said Emerson, "because the same power which sees through [our] eyes, is seen in that spectacle." Being spellbound by the beauty of nature will always pull you out of your separate self, out of the cycle of striving and anxiety and boredom that can so easily wear ruts in our lives. Really, all we have to do is show up, take a breath, quiet our mind.

There was an event back in the sixteenth century involving a group of powerful religious leaders who got together in Rome to discuss a growing crisis in the Catholic Church. The crisis was one of being. It was called "acedia"—best described as a kind of murky, low-level depression. It showed up as a joylessness among the faithful that, for reasons no one could fathom, seemed to be spreading like a cancer through much of Europe. After months trying to figure out the problem with no success, the church fathers are said to have called on that bright and pious spiritual detective Thomas Aquinas. True to his reputation, the future saint spared no effort—praying, fasting, visiting hundreds of churches and thousands of worshippers. After nearly two years Aquinas made his way back to Rome to deliver his findings. People were suffering from acedia, he said, because they'd lost the ability to be in relationship with the beauty of the natural world.

Nineteenth-century British naturalist Richard Jefferies echoed acedia's antidote. The exceeding beauty of the earth, he wrote, "yields a new thought with every petal. The hours when the mind is absorbed by beauty are the only hours when we really live, so that the longer we can stay among these things so much the more is snatched from inevitable Time."

We can certainly end up at arm's length from nature's beauty by virtue of struggling with all that seems to demand our attention—by a life and a world that seems to pulse ever louder and faster. But at the same time I suspect our distance from beauty is furthered by a growing sadness, even guilt, about what's happening to our planet—about the wildfires ravaging our communities, the hurricanes and storm events intensified by the things we humans have done.

We don't much like these feelings. The rush of everyday life makes it all too easy to avoid this sadness and guilt, and at the same time avoid the very relationships with Earth that can help us. This is what psychologists call unprocessed loss, and right now it's calling us to engage in a kind of grief journey. A reconciliation. A heartfelt, intentional walk into the sticky shame we sometimes feel about how we've let our planet down.

But while it's helpful for us to take ownership, to sit for a time with our feelings of regret, there's also this truth: Despite the theatrics of the evening news portraying hurricanes and wildfires as nature's wrath, nature does not loathe humans for the oppression and harm we've leveled against it. Loathing and fury and wrath are finally ours alone. Earth carries on—waxing and waning, ebbing and flowing, warming and flowering.

Beauty is right here, within easy reach. You can feel it under the branches of an old maple tree on a summer afternoon, when in a kind, gentle startle it dawns on you that there's no action for you to take, no problem to solve, no plans to make. Only the shade, the sun, the sound of the breeze in the leaves. And an extraordinary and effortless exchange. You, with every breath out, nurturing the tree. The tree, in turn, giving oxygen for your next breath.

And so the world turns.

And so you turn the world.

# Acknowledgments

Sincerest thanks to my extraordinary agent, Alice Martell, for her shining, steadfast belief in *The Eight Master Lessons of Nature*. From rough idea to final publication, she was both a superb adviser and a dauntless champion. Likewise, endless gratitude to my editor at Dutton Books, John Parsley; thank you, John, for lending your dazzling talents to the tasks of nipping, tucking, recasting, and polishing this manuscript to more clearly reveal its fundamental ideas.

So, too, am I honored to acknowledge a long list of brilliant scientists, historians, and philosophers. In particular, thanks to a wonderful group of biologists, anthropologists, geneticists, physicists, mythologists, ecologists, and psychologists at Harvard and Cambridge Universities, Iowa State, Cornell, and Northern Arizona Universities; also the University of Illinois, the University of Edinburgh, the University of Reading, the University of Sussex, the University of California, Montana State University, the University of Washington, and the University of British Columbia. Thanks, too, to the enormously capable women and men of NASA, the National Oceanic and Atmospheric Administration, and the National Park Service.

Also to psychologist Dr. Eduardo Duran, whose good heart and sharp vision helped me embrace the endless beauty and grace beneath the noise and confusion of daily life.

Finally, I offer my deepest gratitude to the light of my life—my astonishing wife and partner, Dr. Mary M. Clare. During literally thousands of miles spent walking together—on trails and down endless country roads and city sidewalks—she was both spark and nourishment for many of the foundational ideas of this book. Further, as the project drew to a close she was kind enough to offer round after round of deeply thoughtful editing. I stand in awe of Mary's decades of university scholarship in social and developmental psychology and, at the same time, of her boundless passion and skill when it comes to encouraging deep and authentic inquiry in herself and others. She's a blessing to the world.

And to me.

And to these pages.

# About the Author

Gary Ferguson's articles have appeared in a wide variety of national publications. He is also the author of twenty-five books on nature and science, including *The Carry Home: Lessons from the American Wilderness,* which received the 2015 Book of the Year award from the Sigurd Olson Environmental Institute. A frequent guest on National Public Radio, Gary and his wife, Dr. Mary M. Clare, are co-founders of Full Ecology. Find out more at www.fullecology.com.